Acoustic Signals and Hearing

A Time-Envelope and Phase Spectral Approach

Acoustic Signals and Hearing

A Time-Envelope and Phase Spectral Approach

Mikio Tohyama
Research consulting firm: Wave Science Study
Fujisawa-shi, Kanagawa, Japan

Academic Press is an imprint of Elsevier
125 London Wall, London EC2Y 5AS, United Kingdom
525 B Street, Suite 1650, San Diego, CA 92101, United States
50 Hampshire Street, 5th Floor, Cambridge, MA 02139, United States
The Boulevard, Langford Lane, Kidlington, Oxford OX5 1GB, United Kingdom

Library of Congress Cataloging-in-Publication Data
A catalog record for this book is available from the Library of Congress

British Library Cataloguing-in-Publication Data
A catalogue record for this book is available from the British Library

ISBN: 978-0-12-816391-7

For information on all Academic Press publications
visit our website at https://www.elsevier.com/books-and-journals

Publisher: Mara Conner
Acquisitions Editor: Tim Pitts
Editorial Project Manager: John Leonard
Production Project Manager: Nirmala Arumugam
Designer: Christian J. Bilbow

Typeset by VTeX

Contents

List of figures

Biography

Dr. Mikio Tohyama

Mikio Tohyama holds a Doctorate of Engineering from Waseda University, Tokyo, Japan.

Beginning in 1975, and for the next 18 years, Dr. Tohyama was involved in research projects in acoustics, vibration, and signal processing at the Nippon Telegraph and Telephone (NTT) Research Laboratories. He was professor at Kogakuin University (1993–2003) and Waseda University (2003–2012) in sound and auditory perception. Since 2012 he has operated his research consulting firm, Wave Science Study (WSS). His present interest is sound signature analysis oriented toward sound field perception. Dr. Tohyama enjoys playing the piano every day.

e-mail address: m_tohyama@fuji.waseda.jp

home page: https://wavesciencestudy.com

Preface

Acoustical issues of interest to audio engineers or research students range widely from tackling theoretical problems to perceiving everyday auditory events. This book will be helpful for readers interested in selected areas regarding acoustic signals and hearing, from speech intelligibility to spatial hearing.

Sound is an elastic wave that can be represented by mathematical formulations. A sinusoidal function, which is periodic but without any harmonics, is a basis of the mathematical representations of audible sounds. On the other hand, a sinusoidal sound gives the sensation of pitch that corresponds to the frequency of the sinusoidal wave, even if there are no harmonics. The fundamental attributes of the perception of sound are pitch and signal dynamics in the waveform. A sinusoidal wave has no perceptible signal dynamics.

A superposition of sinusoidal functions produces signal dynamics with the different frequencies shown by modulation in the time waveform. The envelope of the modulation is one of the most significant signal signatures from a perceptual standpoint, for example speech intelligibility. Interestingly, the phase relationship between sinusoidal components may largely change the signal dynamics, as developed in Chapter 3 for an informational masker.

The superposition yields spectral dynamics in the time-frequency domain as well, from a frame-wise spectral view. Speech intelligibility is highly sensitive to the narrowband envelopes (such as 1/4–octave band envelopes). A fundamental question arises regarding speech and hearing that is dealt with in Chapter 4: Which is dominant for preserving speech intelligibility, frame-wise magnitude or frame-wise phase spectrum? The time window length of around 30 (ms) that has been conventionally applied for signal analysis suggests the condition for the magnitude spectral dominance. In contrast, the phase could be significant under frame lengths shorter or longer than 30 (ms).

Sound waves transmit from a source to a listener through a sound path (such as a room). Prominent source signatures can be partly preserved in narrowband envelopes, even after traveling through reverberant space, as described in Chapter 7. On the other hand, source signatures of musical sound are conveyed in the spectral envelopes. Vibration of a piano string is a nice example that shows the source signature in both the time and frequency domains, as exhibited in Chapter 5. Readers will be interested in developing the magnitude and phase (or pole–zero) analysis for the source signature estimation.

Signal analysis of the near field to a sound source in a reverberant space is an attractive research area, as described in Chapters 6 and 8. On the other hand, sound quality due to the sound source distance may be intriguing from an audio-engineering perspective such as sound recording. The phase trend in the near field that the direct sound mostly governs could be minimum-phase with the propagation phase. The minimum-phase indicates that the magnitude spectrum makes the near field. Chapter 8 shows that the magnitude-spectral variation due to the sound source distance might be perceptually noticeable.

Why has two-channel stereophony been so widely accepted by the consumers? A key to the stereophonic effect on binaural listening might be the spatial impression (the so-called subjective diffuseness) of the reverberation sound field. The spatial impression could be confirmed by binaural listening to a pair of uncorrelated examples of random noise. Comparing the spatial impression in a reproduced sound field with that for the original field, the merit of two-channel stereophony may be remarkable, as shown in Chapter 9.

If we record a conversation between two people with two microphones, it might be possible to reproduce the perceived separation between them by using two loudspeakers. The separation of sound sources might be another merit of two-channel stereophony. Masking effects of the sound of direct speech on the delayed speech is called the precedence effect, when the delay is within around 30 (ms). In contrast to speech, the precedence effect seems unlikely on a random noise, but a spatial impression might be noticeable by a pair of sounds consisting of a random noise and delayed noise. Envelopes of time waveforms would partly explain the difference between the spatial impression created by a random noise and the precedence effect of speech materials.

Wave propagation is governed by the linear system theory. Sound, however, might be perceived following the signatures of sound sources. This book deals with the sound field from the perspective of signal–signature analysis, for example a time-envelope and phase spectral approach. Some mathematical manipulations are skipped to allow for a smoother reading experience; however, writing out the equations by hand would help the reader follow the subject matter, as developed in the exercises.

Technical terms are mostly listed in the exercises for every chapter, so that the readers might review the information covered. Mathematical expressions are normally expressed in nondimensional forms after normalization, except that the expressions on the physical contents are formulated including the units. The author hopes that the units are informative, and will help the readers follow the equations.

Acknowledgments

The author would like to thank his research colleagues (from NTT Laboratories, Kogakuin University, Waseda University, Hitachi Research Corporation), and Professor Hirofumi Nakajima and Professor Kazunori Miyoshi (Kogakuin University) for their continuous collaboration. The author would also like to thank Hirofumi Onitsuka and his research colleagues (Yamaha Corporation), Hiroyuki Satoh and his research colleagues (Ono Sokki Co. Ltd) for their intensive discussions. The author also acknowledges the very kind guidance provided by Dr. Koji Maruyama (Wolfram Research Asia Ltd) in using Mathematica.

The author expresses his great appreciation to Professor Tammo Houtgast (Amsterdam Free University) and Dr. Yoshimutsu Hirata (SV Research Associates) for their long-term research cooperation and very fruitful discussions. The author sincerely thanks Professor Yoich Ando (Kobe University) for the motivation during the writing of this book. Finally, the author extends his appreciation to all the authors of the research articles referred to in this book.

Mikio Tohyama
September 2019

CHAPTER

Introduction

1

CONTENTS

1.1 Sinusoidal function

A sinusoidal function can represent auditory events rendered by a periodic sound.

1.1.1 Instantaneous magnitude and phase

A sinusoidal function $y(t)$ is defined by the magnitude A and instantaneous phase $\theta(t)$ such that

$$y(t) = A\cos\theta(t) \quad \text{or} \quad y(t) = A\sin\theta(t), \tag{1.1}$$

where $\theta(t)$ is written as

$$\theta(t) = \omega t + \phi. \qquad \text{(rad)} \tag{1.2}$$

The variables ω and ϕ in the instantaneous phase are called the angular frequency (rad/s) and the initial phase (rad). The instantaneous phase $\theta(t)$ is a linear function

Acoustic Signals and Hearing. https://doi.org/10.1016/B978-0-12-816391-7.00009-7

of t. Thus,

$$\frac{d\theta(t)}{dt} = \omega \qquad \text{(rad/s)} \tag{1.3}$$

holds for a sinusoidal function.

The angular frequency ω or $\omega = 2\pi f$ for the frequency $f\,(\mathrm{H_z})$ creates the sensation of pitch by hearing the sound of the wave. The pitch goes higher as the frequency increases; the sensation of pitch is in proportion to the frequency on the logarithmic scale. The frequency interval of a pair of frequencies f_1 and f_2 is called a single octave, when f_2/f_1=2. A musical scale is made by a module in the interval of an octave in general.

Extending the magnitude A into a function of t such that

$$y(t) = A(t)\cos\theta(t), \tag{1.4}$$

then the function $y(t)$ is called a generalized or modulated sinusoidal function, where $A(t)$ is called the instantaneous magnitude or envelope of $y(t)$, and $\cos\theta(t)$ is called the carrier of $y(t)$. When the amplitude (or magnitude) A is a function of time, the waveform is called amplitude modulation. Similarly, the frequency modulation is represented by the carrier with the instantaneous phase $\theta(t)$.

1.1.2 Complex forms for sinusoidal functions

Sinusoidal functions are real functions of real variables. However, a sinusoidal function can be represented by a complex function of real variables.

A complex number is expressed by using a pair of real variables; however, one of the pair of numbers is given by using the imaginary unit i such that

$$z = x + \mathrm{i}y = |z|(\cos\theta + \mathrm{i}\sin\theta), \tag{1.5}$$

$$|z| = \sqrt{x^2 + y^2}, \tag{1.6}$$

where x and y represent the real and imaginary parts of the complex number, respectively.

The imaginary unit i indicates an operation of turning by $\pi/2$ (rad) as shown in Fig. 1.1, where ix is a vector obtained by turning the vector **x** to the left by $\pi/2$. The definition of i as $\mathrm{i}^2 = -1$ would be understood following the operation such that

$$\mathrm{i}(\mathrm{i}\cdot a) = \mathrm{i}^2 a = -a. \tag{1.7}$$

Take a complex number whose absolute is unity. The complex number is located on the unit circle centered at the origin of the complex plane with a radius of unity, as shown in Fig. 1.1. The complex number on the unit circle is defined as

$$z = \cos\theta + \mathrm{i}\sin\theta = \mathrm{e}^{\mathrm{i}\theta} \tag{1.8}$$

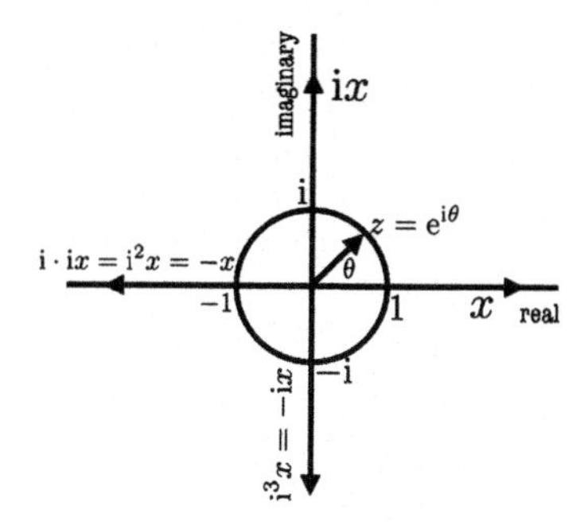

FIGURE 1.1

Unit circle on a complex plane and turning operation by i.

by using the exponential function [1]. The definition given by Eq. (1.8) would be intuitively understood as

$$\cos\theta + \mathrm{i}\sin\theta = \left(\cos\frac{\theta}{n} + \mathrm{i}\sin\frac{\theta}{n}\right)^n \tag{1.9}$$

$$\lim_{n\to\infty}\left(\cos\frac{\theta}{n} + \mathrm{i}\sin\frac{\theta}{n}\right)^n = \lim_{n\to\infty}\left(1 + \frac{\mathrm{i}\theta}{n}\right)^n = \mathrm{e}^{\mathrm{i}\theta}. \tag{1.10}$$

Eq. (1.8) could be extended into the negative θ as

$$\mathrm{e}^{\pm\mathrm{i}\theta} = \cos\theta \pm \mathrm{i}\sin\theta. \tag{1.11}$$

The formula given by Eq. (1.11) leads to

$$\cos\theta = \frac{\mathrm{e}^{\mathrm{i}\theta} + \mathrm{e}^{-\mathrm{i}\theta}}{2} \quad \text{and} \quad \sin\theta = \frac{\mathrm{e}^{\mathrm{i}\theta} - \mathrm{e}^{-\mathrm{i}\theta}}{2\mathrm{i}}. \tag{1.12}$$

A pair of complex variables

$$z = x + \mathrm{i}y = |z|\mathrm{e}^{\mathrm{i}\theta} \quad \text{and} \quad z^* = x - \mathrm{i}y = |z^*|\mathrm{e}^{-\mathrm{i}\theta} \tag{1.13}$$

is denoted by a complex conjugate pair of complex numbers, where

$$|z| = |z^*|, \;\; z + z^* = 2\Re[z], \;\; \text{and} \;\; z - z^* = 2\Im[z]\cdot\mathrm{i}. \tag{1.14}$$

Fig. 1.2 presents examples of locations of complex numbers on the complex plane.

Substituting $\theta = \omega t$ in Eq. (1.12),

$$\cos\omega t = \frac{\mathrm{e}^{\mathrm{i}\omega t} + \mathrm{e}^{-\mathrm{i}\omega t}}{2} \quad \text{and} \quad \sin\omega t = \frac{\mathrm{e}^{\mathrm{i}\omega t} - \mathrm{e}^{-\mathrm{i}\omega t}}{2\mathrm{i}}. \tag{1.15}$$

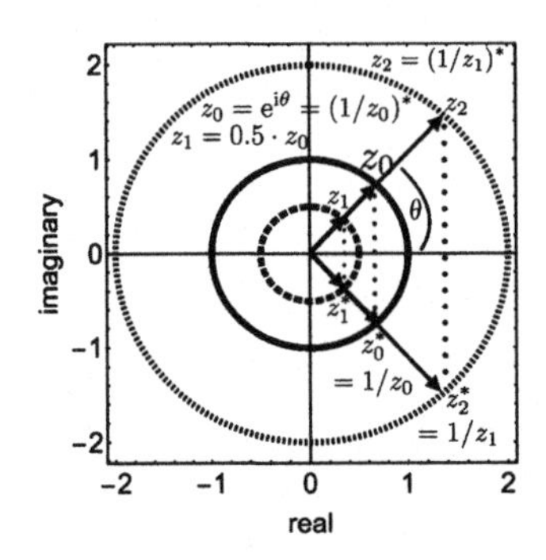

FIGURE 1.2

Complex numbers and their locations on complex plane.

The expressions as complex functions would be helpful to see the periodic nature of sinusoidal functions by the period given by

$$T = \frac{1}{f} = \frac{2\pi}{\omega}. \quad \text{(s)} \tag{1.16}$$

The exponential function with positive ω customary denotes the unit circle following counterclockwise, while the negative ω corresponds to the circle drawn clockwise.

1.1.3 Period and harmonics

A sinusoidal function is periodic where

$$y(x) = y(x + nL) \tag{1.17}$$

holds, L is the period, and n is an integer. Suppose a superposition of sinusoidal functions such as

$$y(x) = \sum_{k=1}^{K} y_k(x) = \sum_{k=1}^{K} A_k \cos(kx + \theta_k). \tag{1.18}$$

The period of $y(x)$ is 2π, which corresponds to the period for

$$y_1(x) = A_1 \cos(x + \theta_1) \tag{1.19}$$

independent of the initial phase θ_1, although the other component

$$y_k(x) = A_k \cos(kx + \theta_k) \tag{1.20}$$

is periodic with the period $2\pi/k$.

The component $y_1(x)$ is called the fundamental and the frequencies of the other components are called the kth harmonics. A superposition of the fundamental and harmonics yields a periodic function. In a periodic function a ratio of the frequencies of the harmonics is a rational number. If the ratio is an irrational number, a superposed function of sinusoidal functions cannot be periodic.

A superposition of sinusoidal components with the fundamental and harmonic frequencies might not be a sinusoidal function any longer, but it could still be periodic. Fig. 1.3 shows examples of periodic and aperiodic functions.

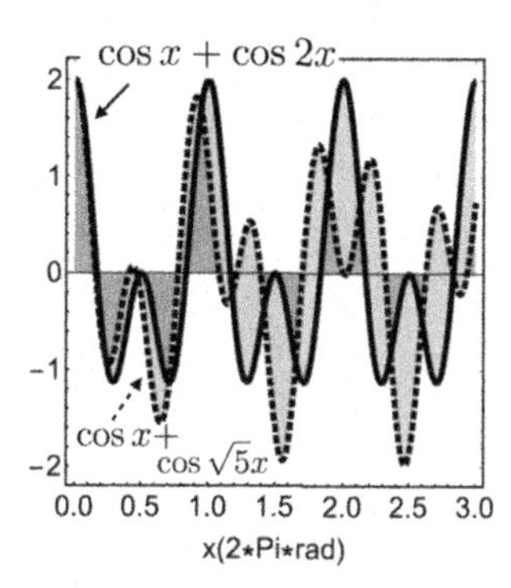

FIGURE 1.3

Periodic and aperiodic functions by superposition of sinusoidal functions.

The function shown by the dotted line is a superposition of sinusoidal functions; however, it is aperiodic because the ratio of the frequencies for the components is an irrational number such that

$$y(x) = \cos x + \cos \sqrt{5}x. \tag{1.21}$$

The sensation of pitch could be possible for sounds and for a single sinusoidal wave (pure tone); it explains that pitch might be basically due to the period of the sound [2]. The frequency of the fundamental corresponds to the period; however, including the fundamental is not necessary in a periodic function. Take the example given by Eq. (1.18). The period of the function is 2π even after removing the fundamental such that

$$\hat{y}(x) = \sum_{k=2}^{K} A_k \cos(kx + \theta_k). \tag{1.22}$$

The sensation of pitch under the condition that the fundamental is removed is called the pitch sensation when missing the fundamental condition [2][3]. Fig. 1.4 illustrates the preservation of the period after removing the fundamental, where $\theta_k = 0$.

The missing fundamental condition could be extended to another case when the harmonics are slightly shifted to different frequencies [4]. Actually, the sound of musical instruments could be composed of the "quasi-harmonics" where the frequencies are not identical to the theoretical frequencies [5]. Take an example of superposed

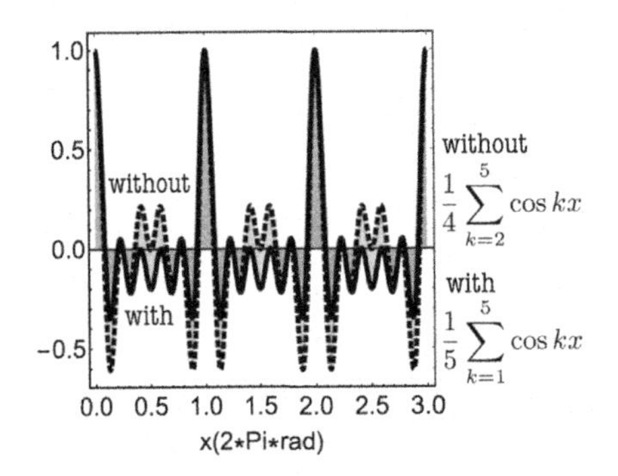

FIGURE 1.4

Sample of a periodic function with or without the fundamental.

sinusoidal components such that

$$y(x) = A_1 \cos k_1 x + A_2 \cos k_2 x \tag{1.23}$$

where $k_2 \cong 2k_1$. The period of function is no longer $2\pi/k_1$ from the theoretical point of view; however, it could still be theoretically periodic as long as k_2/k_1 is a rational number. Fig. 1.5 displays an example where $k_2 = 1.9k_1$.

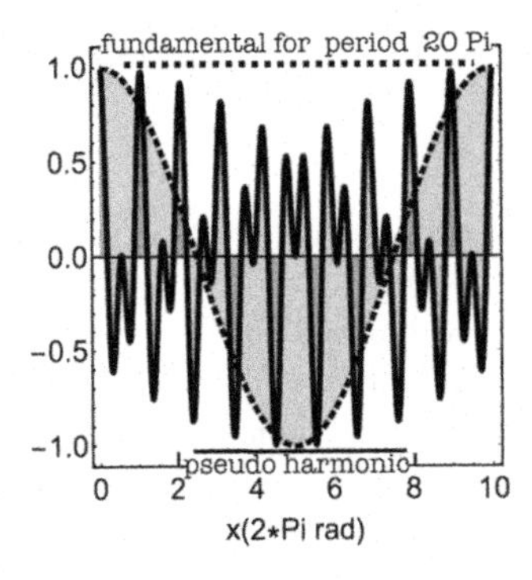

FIGURE 1.5

Example of a pseudo harmonic where the ratio of the second harmonic to the fundamental is 1.9.

The period is 20π because the function can be rewritten as

$$\begin{aligned} y(x) &= \cos x + \cos 1.9x \\ &= \cos 10 \cdot 0.1x + \cos 19 \cdot 0.1x \end{aligned} \tag{1.24}$$

assuming $A_1 = A_2 = 1$, where the fundamental component is

$$y_1(x) = \cos 0.1x \tag{1.25}$$

and the other components are the $10th$ and $19th$ harmonics of the fundamental. This may be another type of missing fundamental condition, where the fundamental could be outside the audible range. Fig. 1.5 represents fluctuations of the envelope within the long period.

1.1.4 Autocorrelation function of periodic function

The autocorrelation function [2][4] is a candidate to confirm the period of a function rather than the waveform itself. Take the example given by Eq. (1.18) once more. The autocorrelation function for the example could be written as

$$r(x) = \frac{1}{\sum_{k=1}^{K} A_k^2} \sum_{k=1}^{K} A_k^2 \cos kx, \tag{1.26}$$

which is again periodic with the period 2π. Here A_k^2, which indicates the square of the magnitude for the kth sinusoidal components, corresponds to the kth power spectral component in terms of waveform analysis [6]. The autocorrelation function depends on the power spectral components independent of the initial phases of the sinusoidal components. The independence of the phase (or sinusoidal or cosinusoidal waveforms) represented by the cosine functions lets the periodic nature be clearly observed rather than the original waveform itself.

The autocorrelation function could be formally extended to aperiodic functions assuming a long-term average. Fig. 1.6 illustrates the autocorrelation functions for the waveforms of periodic and aperiodic functions.

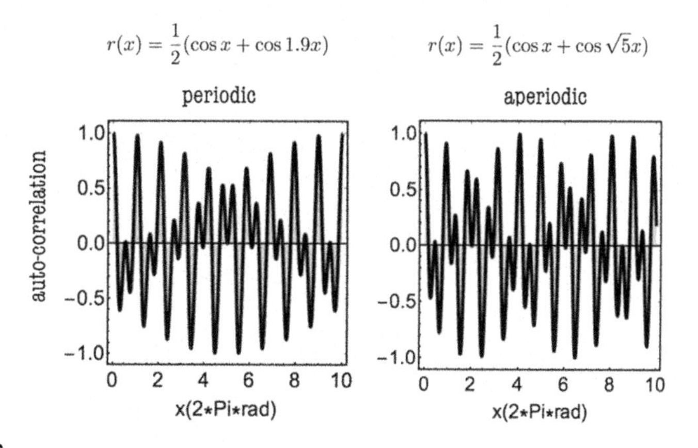

FIGURE 1.6

Autocorrelation functions for periodic *(left)* and aperiodic *(right)* functions.

The period may be not well defined by the autocorrelation function of an aperiodic waveform. On the other hand, the autocorrelation function for a quasi-harmonic waveform could be periodic but with a long period.

1.1.5 Orthogonality of sinusoidal functions and Fourier series expansion

A superposition of sinusoidal functions with different frequencies yields no sinusoidal functions. It implies that a periodic function other than a sinusoidal function might be represented by superposition of harmonic sinusoidal functions.

A Cartesian coordinate system makes a right angle between the pair of horizontal and vertical axes. A concept of the pair making a right angle could be extended into a pair of functions. Suppose a pair of real functions $f_1(x)$ and $f_2(x)$, defined in a domain $0 \leq x \leq L$. The pair of real functions is called an orthogonal pair if

$$\int_0^L f_1(x) f_2(x) dx = 0 \tag{1.27}$$

holds. If the orthogonal relation expressed by Eq. (1.27) holds for the pairs such that

$$\int_0^L f_l(x) f_m(x) dx = \begin{cases} 0 & l \neq m \\ K & l = m, \end{cases} \tag{1.28}$$

the functions $f_l(x)$ lead to an orthogonal system in the domain $0 \leq x \leq L$.

Sinusoidal functions represented by

$$y_l(x) = \sin \frac{l\pi}{L} x \tag{1.29}$$

construct an orthogonal system in the domain $0 \leq x \leq L$, where $l \neq 0$ is an integer, because

$$\frac{2}{L} \int_0^L \sin \frac{l\pi}{L} x \sin \frac{m\pi}{L} x dx = \begin{cases} 0 & l \neq m \\ 1 & l = m. \end{cases} \tag{1.30}$$

The constant shown by $2/L$ is called the normalized factor, thus the sinusoidal function

$$y(x) = \sqrt{\frac{2}{L}} \sin \frac{l\pi}{L} x \tag{1.31}$$

is called a normalized orthogonal function for $0 \leq x \leq L$.

Fig. 1.7 presents an image of decomposition of a vector into orthogonal component vectors. The image of orthogonal decomposition can also be applied for the decomposition of a function into an orthogonal pair of functions. Take the example of a function

$$y(x) = \sin x \cdot \cos 3x = \frac{1}{2} \sin 4x - \frac{1}{2} \sin 2x \tag{1.32}$$

for $0 \leq x \leq 2\pi$. The decomposition above indicates a superposition of the orthogonal pair. The coefficients $1/2$ and $-1/2$ are obtained according to the orthogonality. Set

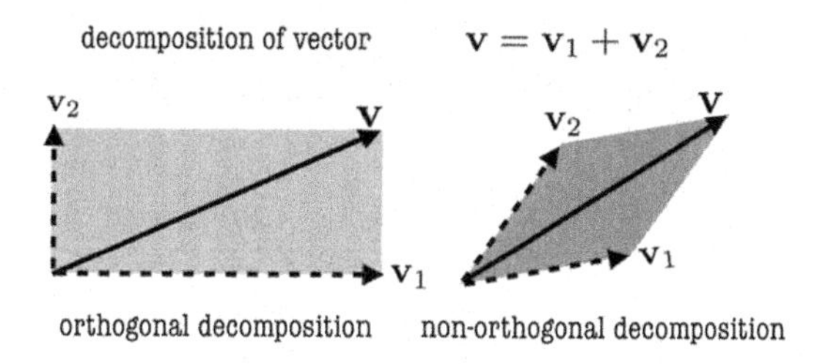

FIGURE 1.7

Decomposition of a vector into an orthogonal or nonorthogonal pair of component vectors.

$y(x)$ given by Eq. (1.32) once more:

$$y(x) = A_1 \sin 4x + A_2 \sin 2x. \tag{1.33}$$

Carry out the integration on the left-hand side,

$$\int_0^{2\pi} y(x) \cdot \sqrt{\frac{1}{\pi}} \sin 4x dx = \frac{\sqrt{\pi}}{2}, \tag{1.34}$$

and similarly on the right-hand side

$$A_1 \int_0^{2\pi} \sqrt{\frac{1}{\pi}} \sin^2 4x dx = A_1 \sqrt{\pi}. \tag{1.35}$$

Consequently, $A_1 = 1/2$ is obtained. For the second term,

$$\int_0^{2\pi} y(x) \cdot \sqrt{\frac{1}{\pi}} \sin 2x dx = -\frac{\sqrt{\pi}}{2} \tag{1.36}$$

is derived from the left-hand side, and

$$A_2 \int_0^{2\pi} \sqrt{\frac{1}{\pi}} \sin^2 2x dx = A_2 \sqrt{\pi} \tag{1.37}$$

from the right-hand side. In addition to A_1, $A_2 = -1/2$ is also obtained, and Eq. (1.32) holds. The integration

$$\int_0^{2\pi} y(x) \cdot \sqrt{\frac{1}{\pi}} \sin 4x dx \text{ and } \int_0^{2\pi} y(x) \cdot \sqrt{\frac{1}{\pi}} \sin 2x dx \tag{1.38}$$

corresponds to the operation of the inner product between a given vector and one of a pair of orthogonal vectors, respectively, as shown in Fig. 1.7.

The mathematical formulation for decomposition of a periodic function into orthogonal sinusoidal functions is called a Fourier series expansion [7], such as

$$f(x) = A_0 + \sum_{k=1}^{N} (A_k \cos kx + B_k \sin kx) \tag{1.39}$$

for $0 \le x \le 2\pi$. Here A_0 denotes the average of the periodic function $f(x)$, and is called a direct current (DC) component in terms of electric engineering. Therefore, B_0 is not necessary for $\sin kx$. The integer k corresponds to the angular frequency of the kth sinusoidal function. Recalling the formula

$$A_k \cos kx + B_k \sin kx = \sqrt{A_k^2 + B_k^2} \cos(kx - \phi_k), \tag{1.40}$$

the magnitude and phase for the kth sinusoidal component would be understood. Both of the coefficients A_k and B_k are necessary to represent a periodic function in general. However, A (or B) is not necessary when the function $f(x)$ is an odd (or even) function. In other words a function can be represented by a superposition of even and odd functions.

The function $f(x)$ should be a smooth function by parts; however, it can be expressed as

$$f(x)|_{x=x_0} = \frac{f(x_0 - 0) + f(x_0 + 0)}{2} \tag{1.41}$$

at the discontinuous point $x = x_0$ in the limit $N \to \infty$. Fig. 1.8 shows examples of a Fourier series expansion where the effects of discontinuity on the convergence of the series can be observed [7]. Truncating the Fourier series at a finite number of terms is called the ideal frequency-band limitation. The effects of the discontinuous function on the convergence of Fourier series expansion, however, can be still observed even after the truncation. Low-pass filtering, instead of truncation, would be necessary to extend a digital audio signal into higher frequency range.

1.2 Response of linear system to sinusoidal input

1.2.1 Linear system

Suppose $x_1(t)$ is the input signal to a system and $y_1(t)$ is the output (or response) of the system. Similarly, take the response as $y_2(t)$ to another input $x_2(t)$. If the response of the system is a sum

$$y(t) = y_1(t) + y_2(t) \tag{1.42}$$

to the input as a sum of

$$x(t) = x_1(t) + x_2(t), \tag{1.43}$$

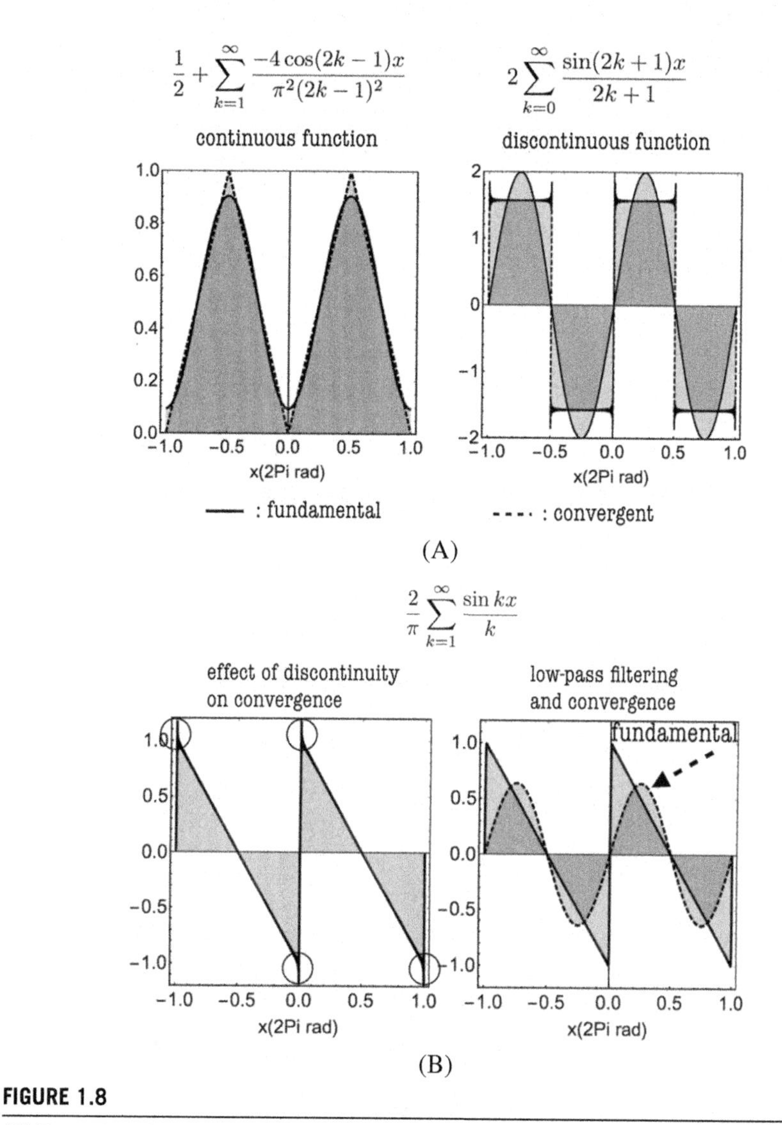

FIGURE 1.8

(A) Examples of convergence of a Fourier series of continuous and discontinuous functions. (B) Smoothing or low-pass filtering for removing effects of discontinuity on convergence.

then the system is a linear system. A sum of the input signals yields the sum of the respective responses.

An example of a linear system is represented by a linear function such as

$$y(x) = ax, \tag{1.44}$$

where x denotes the input and y is the response to the input x. Suppose $x = x_1$, then $y_1 = ax_1$. Similarly, if $x = x_2$, then $y_2 = ax_2$. Consequently, for $x = x_1 + x_2$,

$$y = a(x_1 + x_2) = y_1 + y_2 \tag{1.45}$$

is confirmed.

The linear relationship between the input and output given by Eq. (1.44) could be extended for input functions of a time. Take an input signal $x(t)$, then the response could be written as

$$y(t) = ax(t) \tag{1.46}$$

where a is a constant independent of time. The system defined by Eq. (1.46) is called a linear and time invariant system.

It is not necessary for a linear and time invariant system to be written in the form of Eq. (1.46). For example,

$$y_1(t) = x_1(t) + ax_1(t - \tau), \tag{1.47}$$

$$y_2(t) = x_2(t) + ax_2(t - \tau), \tag{1.48}$$

then

$$y_1(t) + y_2(t) = x_1(t) + x_2(t) + ax_1(t - \tau) + ax_2(t - \tau). \tag{1.49}$$

Eq. (1.49) indicates

$$y(t) = y_1(t) + y_2(t) = x(t) + ax(t - \tau), \tag{1.50}$$

where

$$x(t) = x_1(t) + x_2(t). \tag{1.51}$$

The relation given by Eq. (1.50) shows that the response is composed of the direct and delayed input signals. A superposition of the direct and delayed signals represents a typical response of a linear system, where the number of delayed signals are not necessary to be a single one.

1.2.2 Impulse response and convolution

An ideal input signal is formally defined as an impulse or impulsive signal such that

$$x(t) = \delta(t), \tag{1.52}$$

where $\delta(t)$ denotes the delta function of time. The response of a linear system to an impulsive signal gives the impulse response of the linear system. The impulse response of a linear system can be expressed as a superposition of the direct and delayed signals that can be formulated as a function of time as $h(t)$. The function $h(t)$ shows the density of delayed sounds as a function of time instead of the number of delayed signals.

One of the most fundamental operations of linear systems is convolution. Again, take an input signal $x(t)$ to a linear system that is defined by the impulse response $h(t)$. The response of the system to the input $x(t)$ is formulated by the convolution as [4][6][7]

$$\begin{aligned} y(t) &= \int_0^t h(\tau)x(t-\tau)d\tau = \int_0^t x(\tau)h(t-\tau)d\tau \\ &= x * h(t) = h * x(t), \end{aligned} \tag{1.53}$$

where

$$x(t) = \begin{cases} x(t) & 0 \leq t \\ 0 & t < 0 \end{cases} \quad \text{and} \quad h(t) = \begin{cases} h(t) & 0 \leq t \\ 0 & t < 0. \end{cases} \tag{1.54}$$

Convolution is a commutative operation. The input signal and impulse response, respectively, cannot be identified in the convolution. Reverberant speech is a result of convolution of the original speech signal and the impulse response of the space in which a listener hears the speech. However, daily experience shows that speech may be separately identified from the impulse response of the space or the reverberation. Something different from the linear system theory might be utilized in daily audio communication.

1.2.3 Transient and steady-state responses to sinusoidal input

The response expressed by the convolution shows the output signal as a function of time after the input signal is fed into the linear system. The time dependence of the response can be understood as the transient and steady-state responses. Suppose that the impulse response $h(t)$ is given by

$$h(t) = \mathrm{e}^{-\delta t} \tag{1.55}$$

for $t \geq 0$ assuming $\delta > 0$, which shows an exponential decay curve. The response to a sinusoidal input

$$x(t) = \mathrm{e}^{\mathrm{i}\omega t} \tag{1.56}$$

for $t \geq 0$ becomes

$$\begin{aligned} y(t) &= \int_0^t h(\tau)x(t-\tau)d\tau \\ &= \frac{1}{\delta + \mathrm{i}\omega}\left(1 - \mathrm{e}^{-(\delta+\mathrm{i}\omega)t}\right)\mathrm{e}^{\mathrm{i}\omega t} \to H(\omega)\mathrm{e}^{\mathrm{i}\omega t}, \quad (t \to \infty) \end{aligned} \tag{1.57}$$

where

$$H(\omega) = \frac{1}{\delta + \mathrm{i}\omega}. \tag{1.58}$$

The result of the convolution shows that the response to a sinusoidal input is still the sinusoidal function of time with the identical frequency after the time goes long (or at the steady state), but with a different magnitude and phase expressed as $H(\omega)$. The response that the output signal reaches as time passes long is called the steady-state response. Interestingly, $H(\omega)$, which represents the magnitude and phase at the steady-state response, can be derived:

$$\int_0^\infty h(t)\mathrm{e}^{-\mathrm{i}\omega t}dt = H(\omega). \quad (1.59)$$

The integral formulation given by Eq. (1.59) is called the Fourier transform of the impulse response, which gives the magnitude and phase of the response of a linear system to a sinusoidal input at the steady state. The formulation by the convolution says the frequency of the input signal does not change, even in a reverberant space at the steady state.

The exponentially decaying impulse response approximately represents the reverberation in a room. The decaying constant δ gives the speed of the decay of the impulse response. The reverberation becomes longer (or shorter) as the decaying constant is smaller (or larger). The transient response before reaching the steady-state response becomes quicker (or slower) as the reverberation is shorter (or longer).

In a larger hall the reverberation can be rich, but sometimes musical instruments may not be accurately localized by hearing. This may be true in a listening room as well. Conditions of the recording or reproduction could be basically specified such that

$$D_{T_D} = \frac{\int_0^{T_D} h^2(t)dt}{\int_0^\infty h^2(t)dt}, \quad (1.60)$$

where $T_D \cong 30$ (ms), [8][9][10] where $h(t)$ denotes the impulse response in the recording or reproducing space. The ratio subjectively corresponds to the energy ratio of the direct to the delayed sounds in the superposition constructing the impulse response.

1.3 Sinusoidal sequence

1.3.1 Orthogonality of sinusoidal sequences

Sinusoidal functions are mutually orthogonal in the period as described in Section 1.1.5. A sinusoidal sequence such as

$$x_k(n) = \mathrm{e}^{\mathrm{i}\frac{2\pi}{N}kn} \quad (1.61)$$

is orthogonal to another sinusoidal sequence

$$x_j(n) = \mathrm{e}^{\mathrm{i}\frac{2\pi}{N}jn}, \quad (1.62)$$

where N denotes the period. The orthogonality in the period is written as

$$\sum_{n=0}^{N-1} x_j(n)x_k^*(n) = \sum_{n=0}^{N-1} e^{i\frac{2\pi}{N}(j-k)n} = \begin{cases} 0 & j \neq k \\ N & j = k \end{cases}, \tag{1.63}$$

where $x_k^*(n)$ denotes the complex conjugate of $x_k(n)$.

According to the orthogonality a periodic sequence can be represented as

$$x(n) = \sum_{k=0}^{N-1} X(k)e^{i\frac{2\pi}{N}kn}, \tag{1.64}$$

where

$$X(k) = \frac{1}{N}\sum_{k=0}^{N-1} x(n)e^{-i\frac{2\pi}{N}kn}. \tag{1.65}$$

Eq. (1.64) is called the discrete Fourier expansion of a periodic sequence $x(n)$ and $X(k)$ is the Fourier coefficients of $x(n)$ or the discrete Fourier transform of a periodic sequence of $x(n)$ [7].

The discrete Fourier transform $X(k)$ can be confirmed as

$$\sum_{n=0}^{N-1} x(n)e^{-i\frac{2\pi}{N}kn} = \sum_{n=0}^{N-1}\left(\sum_{l=0}^{N-1} X(l)e^{i\frac{2\pi}{N}ln}\right)e^{-i\frac{2\pi}{N}kn} = N \cdot X(k). \tag{1.66}$$

The periodic sequences $x(n)$ and $X(k)$ can be transformed from one to the other by the discrete Fourier transform according to the orthogonality of the harmonic sinusoidal sequences. The periodic property of $X(k)$ is due to sampling a periodic function $x(t)$ to the periodic sequence $x(n)$. It is a difference from the Fourier series expansion of a periodic function. A sequence whose spectral components are not band-limited could not be sampled without spectral distortion due to the periodic nature of the spectral sequence $X(k)$. If a periodic function is band-limited within the frequency range W_0, the function can be sampled so that the period of spectral sequence might be longer than $2W_0$ [7].

1.3.2 Discrete Fourier transform and filter bank

A periodic sequence $x(n)$ can be represented by using $X(k)$, which denotes the discrete Fourier transform of $x(n)$, and it can be rewritten by using convolution such that

$$x(n) = \sum_{k=0}^{N-1} X(k)e^{i\frac{2\pi}{N}kn} = \frac{1}{N}\sum_{k=0}^{N-1} x_k(n), \tag{1.67}$$

where

$$x_k(n) = \sum_{m=0}^{N-1} x(m)e^{i\frac{2\pi}{N}k(n-m)} = x * h_k(n) \quad \text{and} \quad h_k(n) = e^{i\frac{2\pi}{N}kn}. \tag{1.68}$$

The expression of $x_k(n)$ given by Eq. (1.68) is called the kth sub-band sequence, and $h_k(n)$ denotes the impulse response of the kth sub-band filter.

A set of band-pass (or sub-band) filters is called a filter bank. The filter bank composed of $h_k(n)$ shows a limit case when the impulse response of the sub-band filter is a single sinusoidal sequence. The band widths of the sub-band filters are limits to be 0. According to Eq. (1.68), the sub-band signal $x_k(n)$, which is the response of the kth sub-band filter to the input $x(n)$, is rewritten as

$$x_k(n) = N \cdot X(k)e^{i\frac{2\pi}{N}kn}. \tag{1.69}$$

The magnitude and the phase of the sub-band response, or the sinusoidal sequence, are given by the Fourier transform $X(k)$ such that

$$X(k) = \frac{1}{N}x_k(n)\bigg|_{n=0} = \frac{1}{N}x_k(0). \tag{1.70}$$

Recalling that

$$x(n) = \frac{1}{N}\sum_{k=0}^{N-1} x_k(n) = \sum_{k=0}^{N-1} X(k)e^{i\frac{2\pi}{N}kn}, \tag{1.71}$$

a periodic sequence $x(n)$ can be represented by only using the first entry of every sub-band response as $x_k(0)$. A filter bank is called the perfect reconstruction filter when the sum of the sub-band outputs yields the original sequence $x(n)$. Discrete Fourier transforms make an ideal reconstruction filter.

1.4 Models of sound sources

1.4.1 Symmetric spherical wave

A spherical wave is a representative of sound waves traveling in a medium like the air. The sound pressure wave propagating as a symmetric spherical wave can be written as

$$p(t, r) = \frac{A}{r}e^{i(\omega t - kr)} \quad \text{(Pa)}, \tag{1.72}$$

where r (m) denotes the distance from a source,

$$k = \frac{2\pi}{\lambda} = \frac{\omega}{c} \quad \text{(rad/m)} \tag{1.73}$$

is the wave number in the medium, λ gives the wavelength (m), $\omega = 2\pi f$ denotes the angular frequency (rad/s), f shows the frequency (Hz), and c is the speed of sound in the medium (m/s). Here

$$\lambda = cT = \frac{c}{f} \quad \text{(m)} \tag{1.74}$$

holds and T (s) is the period of the wave. Introducing the propagation time-delay as

$$\tau = \frac{r}{c} \quad \text{(s)}, \tag{1.75}$$

then the symmetric spherical wave can be rewritten as

$$p(t,r) = \frac{A}{r}\mathrm{e}^{\mathrm{i}(\omega t - kr)} = \frac{A}{r}\mathrm{e}^{\mathrm{i}\omega(t-r/c)} = \frac{A}{r}\mathrm{e}^{\mathrm{i}\omega(t-\tau)}. \quad \text{(Pa)} \tag{1.76}$$

The expression in Eq. (1.76) shows that the time-delay τ (s) due to propagation leads to the phase-delay kr (rad):

$$\omega\tau = kr. \quad \text{(rad)} \tag{1.77}$$

1.4.2 Spherical source for spherical wave

A sphere that is uniformly vibrating (from condensation to dilation) is an example of a sound source with symmetric spherical waves. The strength of a spherical source is defined by the volume velocity q (m^3/s) as

$$q = 4\pi a^2 v_s \quad (\mathrm{m}^3/\mathrm{s}), \tag{1.78}$$

where a (m) denotes the radius of the source, and v_s (m/s) gives the vibrating velocity of the surface of the source (or the surface velocity of the source). A spherical wave given by Eq. (1.72) follows the equation of motion for a small portion of a medium as illustrated by Fig. 1.9

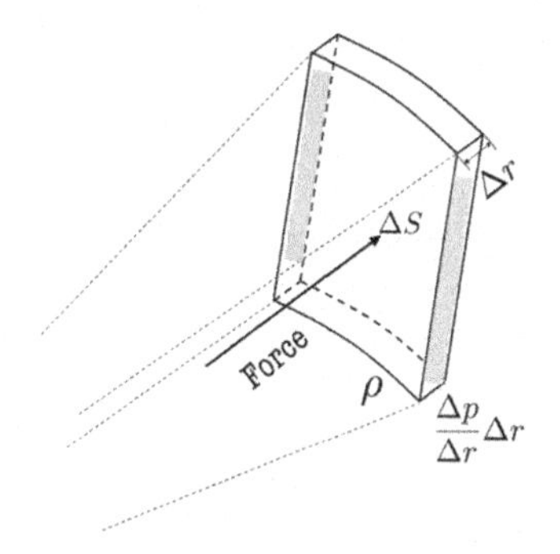

FIGURE 1.9

Image of a small portion of medium in which a spherical wave propagates.

such that

$$\rho \cdot (dS \cdot dr)\frac{\partial v}{\partial t} = -\left(\frac{\partial p}{\partial r} \cdot dr\right) dS. \quad \text{(N)} \tag{1.79}$$

Eq. (1.79) formulates the Newtonian equation of motion of the small portion, where $dS \cdot dr$ gives the volume (m^3) of the small portion, ρ is the volume density of the medium (kg/m^3), v is the vibrating velocity (m/s), and $\partial v/\partial t$ gives the acceleration (m/s^2) for the motion of the small portion. The right-hand side of Eq. (1.79) shows the force applied to the small portion, where p is the sound pressure (Pa); $\partial p/\partial r \cdot dr$ gives the difference in the sound pressure between the two sides; and the negative sign recalls the equation of motion of a mass and spring.

Suppose a spherical source radiates a sinusoidal wave given by Eq. (1.72). An interesting question would be the condition at the surface of the source, or $r = a$. Introducing the strength of the source, then

$$q = 4\pi a^2 v_s = q_0 e^{i\omega t} \quad (m^3/s) \tag{1.80}$$

holds, where

$$v_s = v_0 e^{i\omega t} \quad (m/s) \quad \text{and} \quad q_0 = 4\pi a^2 v_0 \quad (m^3/s), \tag{1.81}$$

and the velocity of the sound wave radiated by the source is imposed on the surface as

$$v = v_s. \quad (m/s) \tag{1.82}$$

Following Eq. (1.79)

$$i\omega\rho v = -\frac{\partial p}{\partial r} \quad (Pa/m) \quad \text{and} \quad v = \frac{p}{i\omega\rho}\frac{1+ikr}{r}. \quad (m/s) \tag{1.83}$$

According to the condition given in Eq. (1.82), the velocity must satisfy

$$v_s = p\left.\frac{1+ikr}{i\omega\rho r}\right|_{r=a} = p\left.\frac{1+ikr}{i\rho ckr}\right|_{r=a}, \quad (m/s) \tag{1.84}$$

where $\omega = ck$ (rad/s). Consequently A in Eq. (1.72) can be written as

$$A = \frac{q_0}{4\pi}\frac{i\rho ck}{1+ika}e^{ika} \quad (Pa \cdot m) \tag{1.85}$$

for the spherical wave radiated by the spherical source. Substituting A given by Eq. (1.85) for Eq. (1.72) yields

$$p(t,r) = \frac{q_0}{4\pi r}\frac{i\rho ck}{1+ika}e^{ika}e^{i(\omega t - kr)}. \quad (Pa) \tag{1.86}$$

The radiated sound pressure depends on ka. Taking a limit, for example $ka \rightarrow$ small, then the sound pressure approaches

$$p(t,r) \rightarrow \mathrm{i}\omega\rho\frac{q_0}{4\pi r}\mathrm{e}^{\mathrm{i}(\omega t-kr)}, \quad \text{(Pa)} \tag{1.87}$$

where the sound source is called a point source, which is the limit of a spherical source when ka becomes small. The magnitude of the sound pressure radiated from a point source increases in proportion to the frequency. A larger source might be necessary to obtain sufficient sound from a source in a lower frequency range.

In contrast, taking a limit when $ka \rightarrow$ large, the radiated sound approaches

$$p(t,r) \rightarrow \frac{q_0}{4\pi r}\frac{\rho c}{a}\mathrm{e}^{\mathrm{i}ka}\mathrm{e}^{\mathrm{i}(\omega t-kr)}. \quad \text{(Pa)} \tag{1.88}$$

The magnitude of the radiated sound pressure can be independent of the frequency in the limit. It confirms that a larger source might be necessary to reproduce a lower frequency range of sound independent of the frequency.

1.4.3 Radiation impedance of spherical source

The phase relationship between the sound pressure and the velocity at the surface of a source characterizes the sound power output of the source [4][7][11]. Taking a limit when $ka \rightarrow$ small, the sound pressure at the surface ($r = a$) approaches

$$p(t,a) \rightarrow \mathrm{i}\omega\rho\frac{q_0}{4\pi a}\mathrm{e}^{\mathrm{i}\omega t} = \rho a\frac{\partial v_s}{\partial t} \quad \text{(Pa)} \tag{1.89}$$

according to Eq. (1.86) where the sound pressure is in proportion to the acceleration of the sound source at the surface.

In contrast, taking a limit when $ka \rightarrow$large, the sound pressure at the surface approaches the value in Eq. (1.88),

$$p(t,a) \rightarrow \rho c v_s, \quad \text{(Pa)} \tag{1.90}$$

where the sound pressure at the surface is in proportion to the surface velocity of the source. The sound pressure and the velocity are in phase at the surface.

The sound power output depends on the phase relationship between the sound pressure and velocity on the surface of the source. The acoustic radiation impedance [11]

$$Z_A = \frac{p(t,a)}{q(t)} = R_A + \mathrm{i}X_A \quad (\mathrm{Pa}\cdot\mathrm{s/m^3}) \tag{1.91}$$

can be derived for a spherical source where

$$R_A = \frac{\rho c}{4\pi a^2}\frac{k^2a^2}{1+k^2a^2} \quad \text{and} \quad X_A = \frac{\rho c}{4\pi a^2}\frac{ka}{1+k^2a^2}. \tag{1.92}$$

The sound power output of a spherical source in a free field is formulated as [11]

$$P = \frac{1}{2}\Re \int_S p \cdot v^* dS = \frac{q_0^2}{2}\Re[Z_A] \quad \text{(W)} \tag{1.93}$$

$$= \frac{q_0^2}{2}\frac{\rho c}{4\pi a^2}\frac{k^2a^2}{1+k^2a^2} \to \begin{cases} \dfrac{q_0^2}{2}\dfrac{\rho c}{4\pi a^2} & ka \to \text{large} \\ \dfrac{q_0^2}{2}\dfrac{\rho c k^2}{4\pi} & ka \to \text{small} \end{cases},$$

where $S(\text{m}^2)$ represents the surface of the source. The sound power output from a spherical source depends on ka, and reduces in proportion to k^2 as ka becomes small.

An example of the effects of environment on the sound power output of a source would be a rigid wall close to the sound source, as shown in Fig. 1.10.

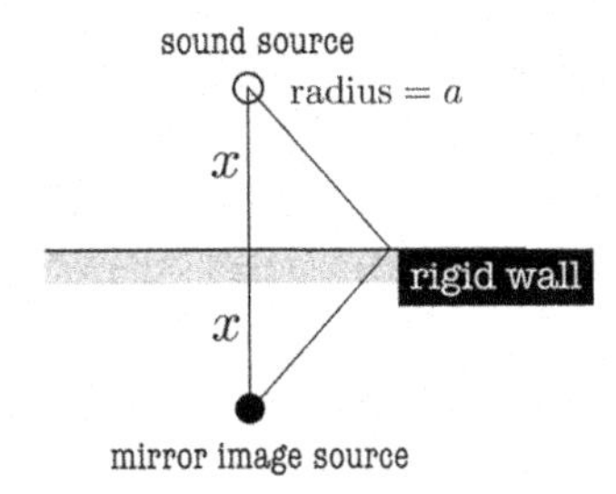

FIGURE 1.10

Small spherical source close to an infinitely extending rigid wall.

The sound pressure created on the surface of the spherical source can be estimated as

$$p(t,a) = \frac{q_0}{4\pi}\frac{\mathrm{i}\rho c k}{\mathrm{i}ka+1}\mathrm{e}^{\mathrm{i}ka}\left(\frac{\mathrm{e}^{-\mathrm{i}ka}}{a} + \frac{\beta}{2x-a}\mathrm{e}^{-\mathrm{i}k(2x-a)}\right)\mathrm{e}^{\mathrm{i}\omega t} \quad \text{(Pa)} \tag{1.94}$$

$$\cong \frac{q_0}{4\pi}\frac{\mathrm{i}\rho c k}{\mathrm{i}ka+1}\mathrm{e}^{\mathrm{i}ka}\left(\frac{\mathrm{e}^{-\mathrm{i}ka}}{a} + \frac{\beta}{2x}\mathrm{e}^{-\mathrm{i}k2x}\right)\mathrm{e}^{\mathrm{i}\omega t},$$

where β denotes the reflection coefficient of the wall. The acoustic radiation impedance can be written as

$$Z_{A_b} = \frac{p(t,a)}{q(t)} \cong \frac{1}{4\pi}\frac{\mathrm{i}\rho c k}{\mathrm{i}ka+1}\mathrm{e}^{\mathrm{i}ka}\left(\frac{\mathrm{e}^{-\mathrm{i}ka}}{a} + \frac{\beta}{2x}\mathrm{e}^{-\mathrm{i}k2x}\right) \quad (\text{Pa}\cdot\text{s/m}^3) \tag{1.95}$$

$$\cong \frac{1}{4\pi}\frac{\mathrm{i}\rho c k}{\mathrm{i}ka+1}\left(\frac{1}{a} + \frac{\beta}{2x}\mathrm{e}^{-\mathrm{i}k2x}\right) = R_{A_b} + \mathrm{i}X_{A_b}$$

when $ka \to$ small. Then the real part of the acoustic radiation impedance approaches

$$R_{A_b} \to R_0\left(1 + \frac{\sin 2kx}{2kx}\right), \quad (\text{Pa}\cdot\text{s/m}^3) \tag{1.96}$$

where

$$R_0 = \frac{\rho c k^2}{4\pi}, \quad (\text{Pa} \cdot \text{s/m}^3) \tag{1.97}$$

which defines the acoustic radiation impedance of a point source as given by the limit of Eq. (1.92) when $ka \rightarrow$small.

Fig. 1.11 illustrates the radiation impedance R_{A_b} when $\beta = 1$. The sound power output, which is in proportion to the real part of the radiation impedance, increases up to twice that in a free field without the rigid wall as the source position comes close to the rigid wall.

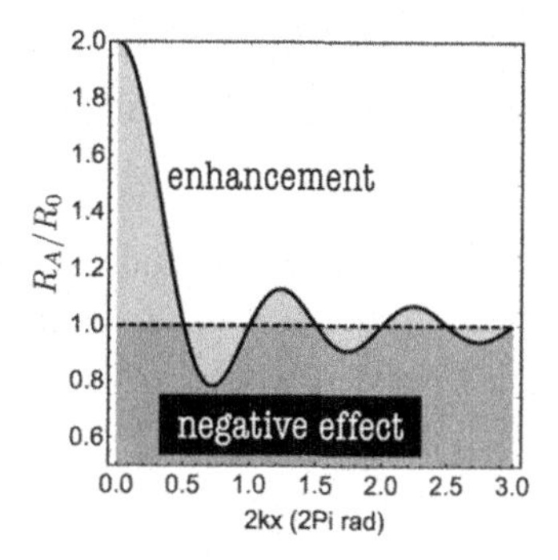

FIGURE 1.11

Acoustic radiation impedance for the real part when the point source is located close to a rigid wall with reflection coefficient $\beta = 1$.

However, the sound power output of a source depends on the source position (defined by the distance from the rigid wall in the example). The sound power possibly decreases more than that for the free field. Recalling that a point source is assumed in the example, a decrease in the sound power output would be understandable due to the position of a source similar in size to a loudspeaker system. A loudspeaker system likely meets the condition under which the sound power output decreases.

Recalling again that the sound power output is formulated as

$$P = \frac{1}{2}\Re\left[\int_S p \cdot v^* dS\right] \quad (\text{W}) \tag{1.98}$$

and assuming the sound pressure to be constant on the surface $S(\text{m}^2)$, then Eq. (1.98) can be rewritten as

$$P = \frac{1}{2}\Re\left[p\int_S v^* dS\right] = \frac{1}{2}\Re[p \cdot q^*], \quad (\text{W}) \tag{1.99}$$

where q denotes the volume velocity (m^3). On the other hand, assuming the vibrating velocity to be constant on the surface, then

$$P = \frac{1}{2}\Re\left[v^* \int_S p dS\right] = \frac{1}{2}\Re[v^* \cdot F], \quad \text{(W)} \tag{1.100}$$

where F denotes the force (N) working on the surface. The radiation impedance is defined in two ways, respectively corresponding to the cases of Eqs. (1.99) and (1.100), such that [11]

$$Z_A = \frac{p}{q} \quad (\text{Pa} \cdot \text{s/m}^3) \tag{1.101}$$

for the acoustic radiation impedance, and

$$Z_M = \frac{F}{v} \quad (\text{N} \cdot \text{s/m}) \tag{1.102}$$

for the mechanical radiation impedance.

The sound power output, or sound power response, of a source is not a function of receiving positions but of the position of the sound source in the environment. Averaging the squared sound pressure through the space in which the sound source is placed, the average is in proportion to the sound power output under the source position [12].

1.5 Exercises

1. Suppose a complex number

$$z = x + \text{i}y = |z|\text{e}^{\text{i}\theta} = r\text{e}^{\text{i}\theta}. \tag{1.103}$$

Show the complex numbers given below on the complex plane:

$$\begin{aligned} &(1)\ z \quad (2)\ z^* \quad (3)\ z^{-1} \\ &(4)\ z^{-1*} \quad (5)\ z + z^* \quad (6)\ z - z^* \end{aligned} \tag{1.104}$$

2. Prove the relationship

$$\begin{aligned} \sin(x \pm y) &= \sin x \cos y \pm \cos x \sin y \\ \cos(x \pm y) &= \cos x \cos y \mp \sin x \sin y \end{aligned} \tag{1.105}$$

by using complex functions for sinusoidal waveforms.

3. Show the fundamental frequencies or the periods for periodic functions below:

$$(1)\ \sin \omega t \quad (2)\ \cos \omega t \quad (3)\ A \sin \omega t + B \cos \omega t \tag{1.106}$$

(4) $\sin(\omega t + \theta)$ (5) $A\sin(\omega t + \theta) + B\cos(\omega t + \phi)$

(6) $A\sin\omega t + B\sin 2\omega t + C\sin 3\omega t$

(7) $A\cos\omega t + B\sin 2\omega t + C\cos 3\omega t$

(8) $A\cos\omega t + B\sin 2.1\omega t$ (9) $\sin 2\omega t\cos\omega t$

(10) $\sin\frac{4.1}{2}\omega t\cos\frac{2.1}{2}\omega t$

4. Obtain the autocorrelation functions for the periodic signals given in 3. The autocorrelation function can be formally defined as

$$r(\tau) = \lim_{T_L\to\infty}\frac{1}{2T_L}\int_{-T_L}^{T_L} f(t)f(t-\tau)dt, \tag{1.107}$$

assuming the long-term average to be 0 for aperiodic functions.

5. Confirm Eq. (1.30).

6. Derive the Fourier series expansions in Fig. 1.8.

7. Confirm that the sequence $X(k)$, which is the discrete Fourier transform of a periodic sequence $x(n)$, $(n = 0, 1, 2, \cdots, N-1)$, is periodic.

8. Derive Eqs. (1.83), (1.85), and (1.86).

9. Confirm and sketch the outline of Eq. (1.92). Where is the cross-point for the real and imaginary parts?

10. Derive Eq. (1.97).

11. Explain the terminologies listed below:
(1) magnitude (2) angular frequency
(3) initial phase (4) octave (5) instantaneous phase
(6) envelope (7) carrier
(8) fundamental (9) harmonics (10) period
(11) periodic function (12) missing fundamental (13) orthogonal functions
(14) Fourier series expansion (15) linear system (16) impulse response
(17) convolution (18) Fourier transform (19) discrete Fourier transform
(20) spherical wave (21) wave length (22) wave number
(23) volume velocity of source (24) speed of sound
(25) sound power output of source (26) radiation impedance of spherical source
(27) mechanical radiation impedance
(28) acoustic radiation impedance

References

[1] S. Lang, Complex Analysis, Springer New-York, Inc., 1999.
[2] R. Meddis, L. O'Mard, A unitary model of pitch perception, J. Acoust. Soc. Am. 102 (3) (1997) 1811–1820.
[3] B.C.J. Moore, An Introduction to the Psychology of Hearing, Academic Press, 1997.
[4] M. Tohyama, Sound in the Time Domain, Springer, 2017.
[5] T. Hasegawa, M. Tohyama, Analysis of spectral and temporal waveforms of piano-string vibration, J. Audio Eng. Soc. 60 (4) (2012) 237–245.
[6] M. Tohyama, Waveform Analysis of Sound, Springer, 2015.
[7] M. Tohyama, T. Koike, Fundamentals of Acoustic Signal Processing, Academic Press, 1998.
[8] R. Thiele, Richtungsverteilung und Zeitfolge der Schallrueckwuerfe in Raeumen, Acustica 3 (1953) 291–302.
[9] T.J. Schlutz, Acoustics of the concert hall, IEEE Spectr. (June 1965) 56–67.
[10] Y. Hirata, Reverberation time of listening room and the definition of reproduced sound, Acustica 41 (3) (1978) 222–224.
[11] M. Tohyama, Sound and Signals, Springer, 2011.
[12] M. Tohyama, A. Imai, H. Tachibana, The relative variance in sound power measurements using reverberation rooms, J. Sound Vib. 128 (1) (1989) 57–69.

CHAPTER

Resonance systems

2

CONTENTS

2.1 Single resonator

2.1.1 Single-degree-of-freedom system

The vibration of a single mass with an elastic spring shown in the left panel of Fig. 2.1 is called a single-degree-of-freedom system because only one eigen angular frequency or frequency of the free vibration

$$\omega_0 = \sqrt{K/M} \quad \text{(rad/s)} \tag{2.1}$$

is found, where M (kg) and K (N/m) are the mass and the spring constant, respectively.

Acoustic Signals and Hearing. https://doi.org/10.1016/B978-0-12-816391-7.00010-3

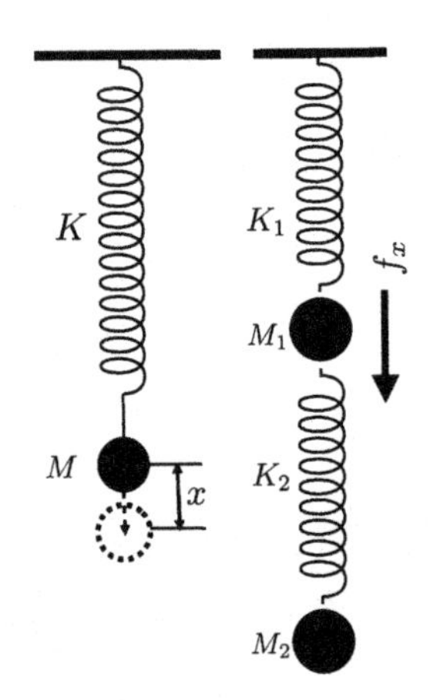

FIGURE 2.1

Single-degree-of-freedom (left) and two-degree-of-freedom (right) vibrating systems.

The eigen angular frequency can be defined as the frequency of the free vibration that follows the equation of the motion of the mass such that

$$M\frac{d^2}{dt^2}x(t) + Kx(t) = 0 \quad \text{(N)} \tag{2.2}$$

under the negligibly small damping constant, where $x(t)$ gives the displacement (m) of the mass.

A solution of Eq. (2.2) is given by [1]

$$x(t) = A\cos(\omega_0 t + \theta), \quad \text{(m)} \tag{2.3}$$

where ω_0 (rad/s) denotes the eigen angular frequency, and A (m) and θ (rad) are determined by the initial conditions of the vibration $x(t)$. The potential and kinetic energy of the displacement $x(t)$ can be derived as [1]

$$E_p = \frac{1}{2}Kx^2(t) \quad \text{(J)} \tag{2.4}$$

$$E_k = \frac{1}{2}Mv^2(t) = \frac{1}{2}M\left(\frac{dx(t)}{dt}\right)^2, \tag{2.5}$$

where $v(t)$ (m/s) denotes the vibrating velocity. Substituting Eq. (2.3) for $x(t)$, the potential and kinetic energy are expressed as

$$E_p = \frac{1}{2}KA^2\cos^2(\omega t + \theta) \quad \text{(J)} \tag{2.6}$$

$$E_k = \frac{1}{2}MA^2\omega^2\sin^2(\omega t + \theta). \tag{2.7}$$

For the eigenmotion with the eigenfrequency such that $\omega = \omega_{00}$

$$E_p + E_k = \frac{1}{2} K A^2 \quad \text{(J)}, \tag{2.8}$$

the energy preservation law holds in the vibration of the eigenfrequency.

2.1.2 Loudspeaker cabinet as a single-degree-of-freedom system

A conceptual model of a loudspeaker unit in a cabinet can be represented by a single-degree-of-freedom system, as shown in Fig. 2.2.

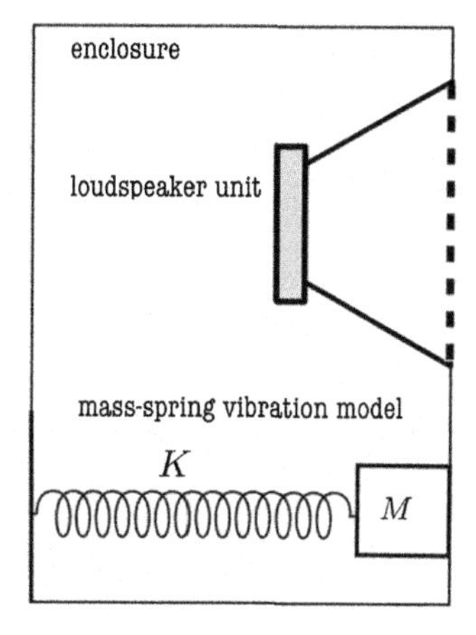

FIGURE 2.2

Loudspeaker unit in an enclosure as a single-degree-of-freedom system.

The sound pressure inside the cabinet becomes higher as the diaphragm of the loudspeaker unit moves inside (in the negative direction) because the air inside the enclosure is condensed. The air inside the enclosure dilates so that the air pressure might decrease, if the diaphragm moves outside the enclosure (in the positive direction). The variation in the sound pressure inside the enclosure and the movement of the diaphragm are in opposite directions or in antiphase.

In contrast, the pressure (outside the enclosure) in front of the diaphragm results from the acceleration of the diaphragm following the Newtonian law of motion. It indicates that the variations in the sound pressure are in phase inside and outside the enclosure [1].

2.2 Coupled oscillator

2.2.1 Two-degree-of-freedom system

The right panel of Fig. 2.1 is an example of a two-degree-of-freedom vibration system. The coupling oscillator would be a fundamental model for the reduction of machinery vibration [2]. Suppose that the external force, $F e^{i\omega t}$ (N), works on a mass

M_1. The motion of the coupled masses follows a pair of equations such that [3]

$$M_1 \frac{d^2}{dt^2} x_1(t) = -K_1 x_1(t) - K_2(x_1(t) - x_2(t)) + F e^{i\omega t} \quad \text{(N)} \tag{2.9}$$

$$M_2 \frac{d^2}{dt^2} x_2(t) = -K_2(x_2(t) - x_1(t)). \tag{2.10}$$

Substituting

$$x_1(t) = X_1(\omega) e^{i\omega t} \quad \text{and} \quad x_2(t) = X_2(\omega) e^{i\omega t} \quad \text{(m)} \tag{2.11}$$

into the simultaneous equations, then the transmission ratio T_{12} is obtained such that

$$T_{12} = \frac{X_2(\omega)}{X_1(\omega)} = \frac{\omega_2^2}{\omega_2^2 - \omega^2}, \tag{2.12}$$

where $\omega_2^2 = K_2/M_2$.

The ratio T_{12} may indicate that the vibration of the frequency, close to the resonance frequency ω_2 of the second component, transmits largely, so that the vibration of component 1 might almost stop. In contrast, the vibration mostly remains at component 1 without transmission when the frequency of the vibration is far from the resonance frequency of the second component [1]. The phase of the vibration is intriguing. The transmission ratio shows the change in the phase relationship between the vibrations of the first and second components.

2.2.2 Coupled oscillator modeling for bass-reflection loudspeaker system

A bass-reflection type of loudspeaker system is popular in audio equipment. Fig. 2.3 is a conceptual model of bass-reflection loudspeaker system where a two-degree-of-freedom vibration system is fitted. The phase relationship between the two vibrating components is in phase for $\omega < \omega_2$, but in antiphase for $\omega > \omega_2$, where $\omega_2 = \sqrt{K_2/M_2}$ (M_2 denotes the mass of the air in the duct and K_2 is determined by the volume of the air in the cabinet), and $\omega_1 = \sqrt{K_1/M_1}$ (M_1 is the mass of the diaphragm and K_1 gives the spring constant for holding the diaphragm). The angular frequency ω_2 is the Helmholtz resonance frequency [1].

The phase relationship is interesting from the point of view of the effect of bass-reflection on the radiated sound. The sound pressure is in phase between the inside and outside the enclosure, as shown in Fig. 2.2. On the other hand, the vibrations are in antiphase between the first and second components for $\omega > \omega_2$, as stated in Eq. (2.12). This antiphase pair of vibrations for the loudspeaker diaphragm and air duct makes an in-phase pair of the motions as a pair of sound sources that may increase the radiation of the sound from the loudspeaker system.

Fig. 2.4 illustrates the sound pressure level and phase differences, corresponding to $1/T_{12}$ in Eq. (2.12), between A (in front of the diaphragm) and B (the air duct), as shown in Fig. 2.3 [4].

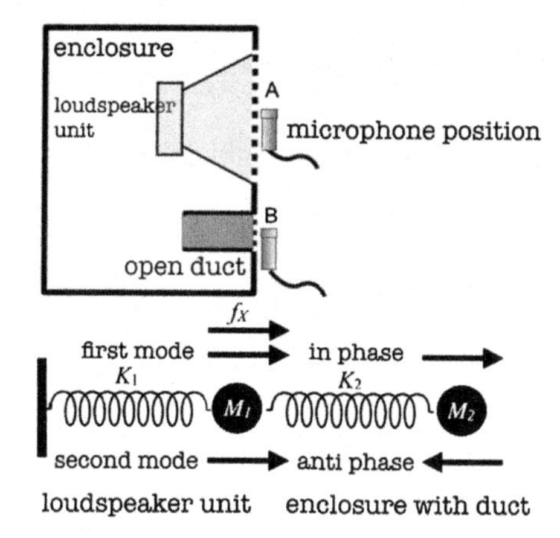

FIGURE 2.3

Two-degree-of-freedom system modeling for a bass-reflection type of loudspeaker system.

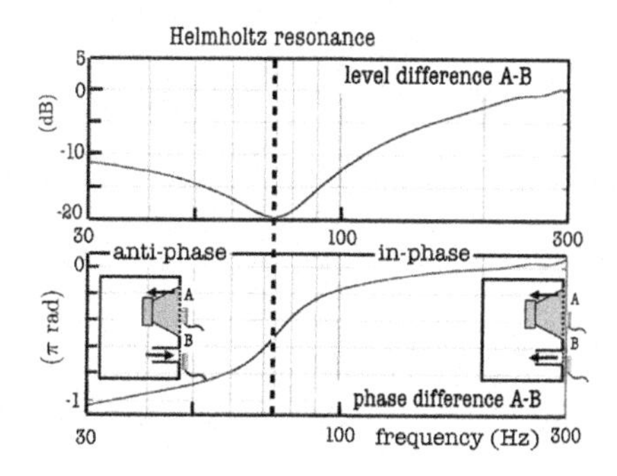

FIGURE 2.4

Level and phase differences in radiated sound between A and B in Fig. 2.3, corresponding to $1/T_{12}$ in Eq. (2.12).

The transmission of the vibration of the diaphragm to the air duct is confirmed as the frequency increases and passes the Helmholtz resonance, in which $A/B \to 0$. The phase difference confirms the in-phase relationship in the sound pressure variations between the positions of the diaphragm and the air duct in the higher frequency range than the Helmholtz resonance. Enforcement in the radiated sound pressure from the loudspeaker unit may be possible in the higher frequency range than the Helmholtz resonance because of the in-phase superposition of the radiated sound from the air duct.

2.2.3 Zeros and transfer function

The transmission ratio given by Eq. (2.12) can be rewritten as

$$H(\omega) = \frac{1}{T_{12}} = \frac{\omega_2^2 - \omega^2}{\omega_2^2}, \tag{2.13}$$

which is a function of ω. Extending the angular frequency into the complex frequency domain such that [5]

$$\omega^c = \omega + \mathrm{i}\delta, \tag{2.14}$$

the transmission ratio could be extended into the transfer function of ω^c. Fig. 2.5 illustrates the complex frequency plane, where δ denotes the distance from the real frequency axis ω.

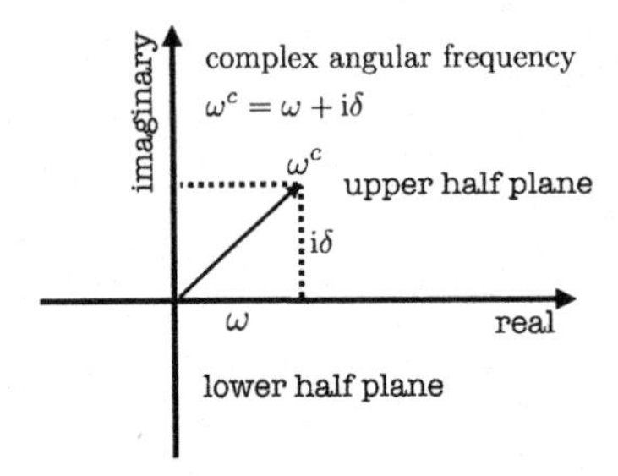

FIGURE 2.5

Complex frequency plane.

Substituting the complex angular frequency ω^c for ω, then $H(\omega^c)$ becomes a complex function.

The complex frequency indicates the frequency of a decaying vibration. Introducing a decaying constant given by δ into the free vibration, then the frequency ω_2 can be expressed as

$$\omega_2^c = \omega_2 + \mathrm{i}\delta_2. \tag{2.15}$$

The magnitude of the transfer function, when the real angular frequency ω increases over ω_2 on the real frequency axis close to ω_2^c in the complex frequency plane, becomes minimum when $\omega = \omega_2$; however, the magnitude is not zero. The real angular frequency $\omega = \omega_2$ does not equal zero:

$$H(\omega^c)\big|_{\omega^c = \omega_2} \neq 0. \tag{2.16}$$

The complex (in general) angular frequency ω^c that satisfies

$$H(\omega^c) = 0 \tag{2.17}$$

is called the zero of the transfer function. In the example shown by Fig. 2.4, ω^c is not on the real axis, but in a complex plane. The vibration of the component M_1 (diaphragm) transmits mostly to M_2 (air duct) at the frequency, and the magnitude is at a minimum. Consequently, there is almost no vibration of the diaphragm, although it does not stop completely because the minimum is not zero.

The phase response clearly makes a change, when the angular frequency increases over ω_{z_1} or ω_{z_2} (shown Fig. 2.6) on the real frequency axis. Fig. 2.6 is an image of the location of the zero and the phase change in the complex plane.

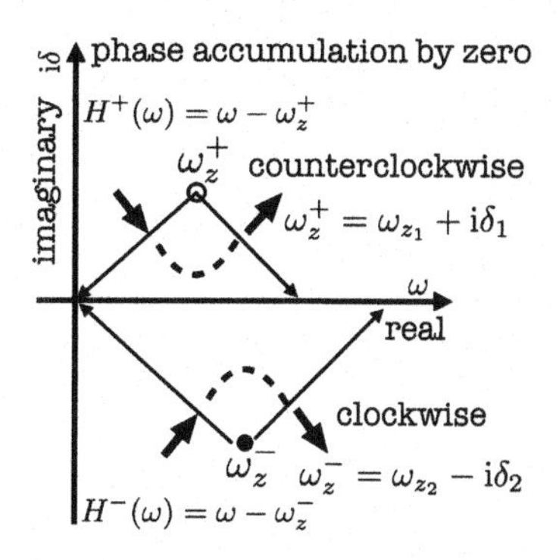

FIGURE 2.6

Zeros in the complex frequency plane and phase change.

The movement of the phase angle is made as the angular frequency goes through on the real frequency axis. The positive phase change (counterclockwise), as shown in Fig. 2.4, indicates that the zero is located above the real frequency axis (in the upper half-plane) [6].

2.3 Response to sinusoidal input

2.3.1 Steady-state response and poles of transfer function

Suppose a single-degree-of-freedom system excited by a sinusoidal function such that

$$M\frac{d^2}{dt^2}x(t) + R\frac{d}{dt}x(t) + Kx(t) = F_x e^{i\omega t}, \quad \text{(N)} \tag{2.18}$$

where the damping effect by R (N · s/m) is introduced. Assuming the response to be

$$x(t) = X(\omega)e^{i\omega t} \quad \text{(m)} \tag{2.19}$$

at the steady state [1][5], then

$$X(\omega) = \frac{F_x/M}{\omega_0^2 - \omega^2 + i\frac{R}{M}\omega} = \frac{-F_x/M}{(\omega - \omega_p^-)(\omega - \omega_p^+)} \quad \text{(m)} \tag{2.20}$$

is derived, where

$$\omega_p^+ = \frac{R}{2M}i + \omega_d, \quad \omega_p^- = \frac{R}{2M}i - \omega_d, \quad \text{and} \quad \omega_d^2 = \omega_0^2 - \left(\frac{R}{2M}\right)^2. \tag{2.21}$$

Taking the ratio between the exciting force and the displacement of the mass

$$G(\omega) = \frac{X(\omega)}{F_x/M} = \frac{-1}{(\omega - \omega_p^-)(\omega - \omega_p^+)}, \quad (\mathrm{s}^2) \tag{2.22}$$

where ω_p^- and ω_p^+ are singularities or the poles of $G(\omega)$. The poles are symmetrically located in the upper half-plane with respect to the imaginary axis, as shown in Fig. 2.7.

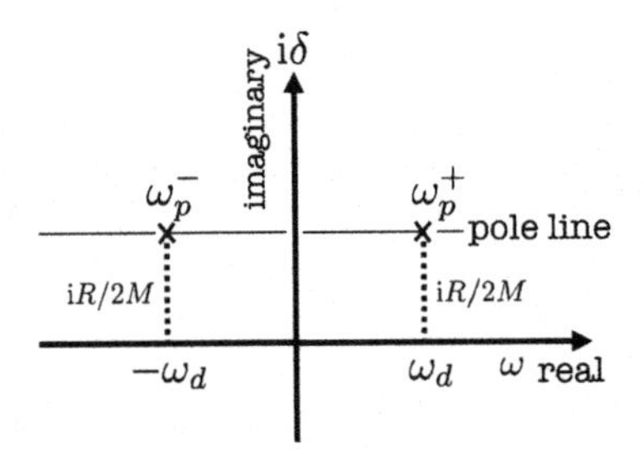

FIGURE 2.7

Symmetric pair of poles with respect to imaginary axis in the upper half of the complex frequency plane.

The distance between the poles and the real frequency axis becomes greater as the damping factor $R/2M$ (1/s) becomes larger.

2.3.2 Magnitude response and resonance

Returning to Eq. (2.20) and taking square of the magnitude of $X(\omega)$, then

$$|X(\omega)|^2 = \frac{(F_x/M)^2}{(\omega_0^2 - \omega^2)^2 + (\frac{R}{M})^2\omega^2} \quad (\mathrm{m}^2) \tag{2.23}$$

which reaches its maximum at

$$\omega_{res} = \sqrt{\omega_0^2 - \frac{1}{2}\left(\frac{R}{M}\right)^2}. \quad (\mathrm{rad/s}) \tag{2.24}$$

The angular frequency ω_{res} at which the magnitude is at its maximum is called the resonance angular frequency. The resonance angular frequency is a little lower than the value of ω_d given by Eq. (2.21), which is the magnitude of the real part of the pole ω_{p_1} or ω_{p_2}, and becomes lower as the damping increases [1][7].

2.3.3 Frequency selectivity by resonance

Fig. 2.8 shows examples of the squared magnitude of the resonances under damping conditions.

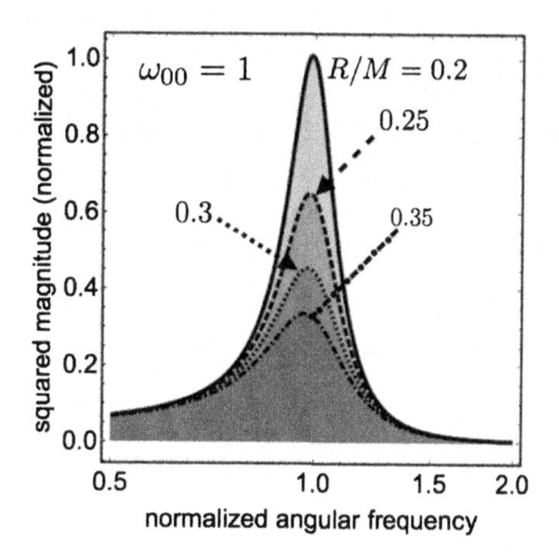

FIGURE 2.8

Examples of resonances under damping conditions where ω_{00} indicates normalized ω_0.

The frequency selectivity according to the resonance becomes broader as the damping increases.

The frequency selectivity for hearing is very narrow. The barely noticeable frequency difference could be less than 3 (Hz) at a level of around 1000 (Hz) [8][9]. Assuming that the limit of detection in the sound magnitude difference is around 1 (dB) [9], a close-up of the 1 (dB) range around the resonance peak implies a very narrow frequency selectivity, as shown in Fig. 2.9.

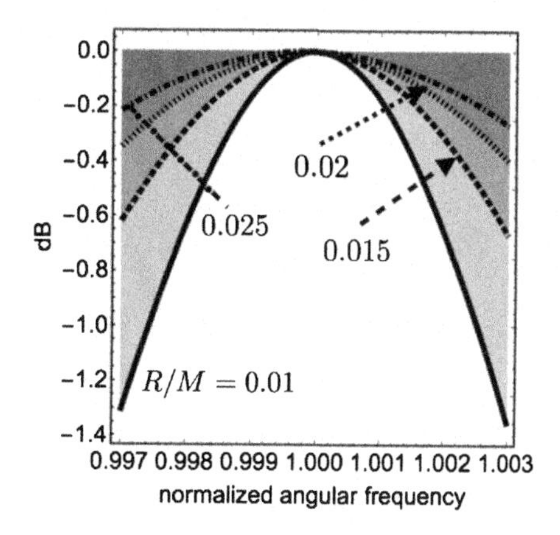

FIGURE 2.9

Close-up of resonance peaks under damping conditions.

Interestingly, active mechanisms including a feedback loop have been investigated in the cochlea dynamics [10] instead of passive resonance for frequency responses of the basilar membrane. Another approach to frequency selectivity is suggested in

Reference [11], too. Suppose the frequency of a pure tone is identical to the center frequency of the central filter of the three sub-band filters. Compare the outputs of the three filters (left, central, and right), the outputs of the pair of side filters (left and right) might be almost the same if the response of every filter is identical and symmetric with respect to the center frequency. A slight variation in the frequency of the pure tone makes a corresponding change in the output of the central filter. The output of the central filter, for example, may show only a small variation around 0.4 (dB) in the example of Fig. 2.9. However, the balance of responses between the pair of side filters may change due to the off-central frequency. The off-central imbalance between the side filters, due to the slight change of the pure-tone frequency, might be helpful to detect the slight variation in the frequency [11].

2.3.4 Impulse and transient responses

The equation of motion of a single-degree-of-freedom system can be written as

$$\frac{d^2}{dt^2}x(t) + \frac{R}{M}\frac{d}{dt}x(t) + \frac{K}{M}x(t) = G_0\delta(t) \quad (\mathrm{m/s^2}) \tag{2.25}$$

under impulsive excitation, where

$$f(t) = G_0\delta(t). \quad (\mathrm{m/s^2}) \tag{2.26}$$

Substituting

$$x(t) = \frac{1}{2\pi}\int_{-\infty}^{\infty} \hat{X}(\omega)\mathrm{e}^{\mathrm{i}\omega t}\,d\omega \quad (\mathrm{m}) \tag{2.27}$$

$$G_0\delta(t) = G_0\frac{1}{2\pi}\int_{-\infty}^{\infty} \mathrm{e}^{\mathrm{i}\omega t}\,d\omega \quad (\mathrm{m/s^2}) \tag{2.28}$$

by the Fourier transforms, then

$$\hat{X}(\omega) = \frac{-G_0}{(\omega - \omega_{p_1})(\omega - \omega_{p_2})}, \quad (\mathrm{m \cdot s}) \tag{2.29}$$

which corresponds to $X(\omega)$ given by Eq. (2.20), where $\omega_{p_1} = \omega_p^+$, $\omega_{p_2} = \omega_p^-$.

The solution of Eq. (2.25), $x(t)$, is called the impulse response (customarily denoted $h(t)$), and its Fourier transform in Eq. (2.29) (or $H(\omega)$) is the response to a sinusoidal wave $\mathrm{e}^{\mathrm{i}\omega t}$ at the steady state. The impulse response $h(t)$ is expressed as [5]

$$x(t) = h(t) = \frac{1}{2\pi}\int_{-\infty}^{\infty} \hat{X}(\omega)\mathrm{e}^{\mathrm{i}\omega t}\,d\omega = G_0\frac{\sin\omega_d t}{\omega_d}\mathrm{e}^{-\delta_0 t}. \quad (\mathrm{m}) \tag{2.30}$$

The impulse response indicates decaying oscillation with the frequency of ω_d.

The response $y(t)$ to a sinusoidal input $x(t) = e^{i\omega t}$ into a linear system is written as

$$y(t) = x * h(t) = e^{i\omega t} \cdot H(\omega, t) = e^{i\omega t} \int_0^t h(\tau) e^{-i\omega\tau} d\tau. \tag{2.31}$$

Fig. 2.10 shows examples of the magnitude frequency characteristics of the responses to the sinusoidal input.

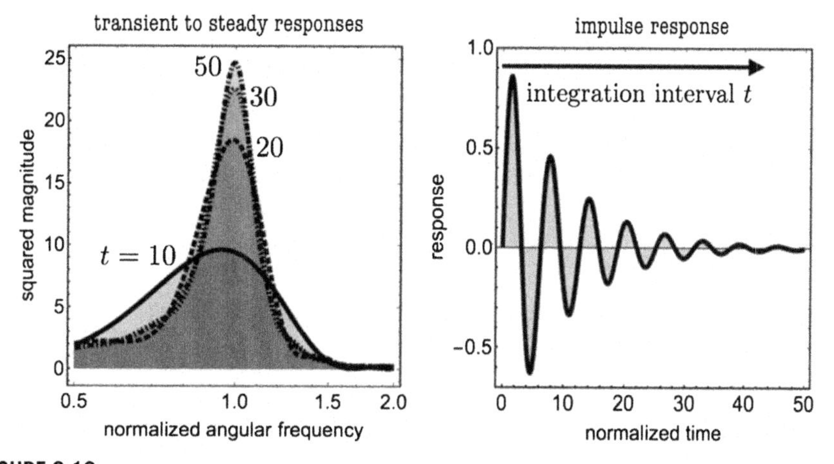

FIGURE 2.10

Squared-magnitude frequency characteristics for $H(\omega, t)$ *(left)* and impulse response *(right)*.

The resonance becomes prominent as the length of the impulse response record becomes longer. On the other hand, the transient response takes a longer time to reach the steady-state response for a longer impulse response. A quick response to an input signal might be realized for a damped system rather than a highly resonant system. Resonance with narrower frequency selectivity yields slower responses, while damped oscillation systems with wider frequency selectivity yields quicker responses to an input signal.

2.4 Wave equation and sound propagation in a one-dimensional system

2.4.1 Wave equation

Vibration of a mass and spring system is characterized by the eigenfrequency or the frequency of free vibrations. Propagation of vibration as a wave can be observed for a series of spring–mass components, where the speed of sound is the fundamental concept instead of the eigenfrequencies for spring–mass systems [1][12]. The Newtonian equation of motion is written as

$$M\frac{d^2u_i(t)}{dt^2} = f_i \quad \text{(N)} \tag{2.32}$$

for the ith element where u_i (m) gives the displacement (or stretch) of the ith spring and f_i (N) indicates the force working on the mass attached to the ith spring. The restoring force f_i against the stretch of the ith spring is written as

$$f_i = K(u_{i+1} - u_i) - K(u_i - u_{i-1}), \quad \text{(N)} \tag{2.33}$$

where K (N/m) is the spring constant.

Assuming a limit case when the spring–mass components are densely connected in the series so that u_i and f_i might be smooth functions of time t and place x, then the wave equation can be derived for one-dimensional space as [1][12]

$$\frac{\partial^2 u(x,t)}{\partial t^2} = c^2\frac{\partial^2 u(x,t)}{\partial x^2}, \quad (\text{m/s}^2) \tag{2.34}$$

where c (m/s) denotes the speed of sound. Eq. (2.34) is called the wave equation for one-dimensional space.

2.4.2 General solution of wave equation

The wave equation in one-dimensional space has a general solution, such as [7][13]

$$u(x,t) = f(ct - x) + g(ct + x), \quad \text{(m)} \tag{2.35}$$

which satisfies the wave equation in a one-dimensional field without boundaries. The first (or second) term indicates the progressive wave toward the positive (or negative) direction of x. The functions f and g represent the waveforms for the progressive waves that are represented as smoothing (or differentiable) functions. The variable c (m/s) in the wave equation indicates the speed of sound traveling in the medium. Fig. 2.11 shows an example of traveling waves by setting

$$f(z) = g(z) = Ae^{-az^2}, \tag{2.36}$$

where $z = ct \pm x$.

2.4.3 Traveling of sinusoidal wave

Suppose that

$$f(z) = g(z) = Ae^{ikz}, \tag{2.37}$$

where $z = ct \pm x$. The traveling wave toward the positive direction of x is expressed as

$$f(ct - x) = Ae^{ik(ct-x)} = Ae^{i(\omega t - kx)}, \tag{2.38}$$

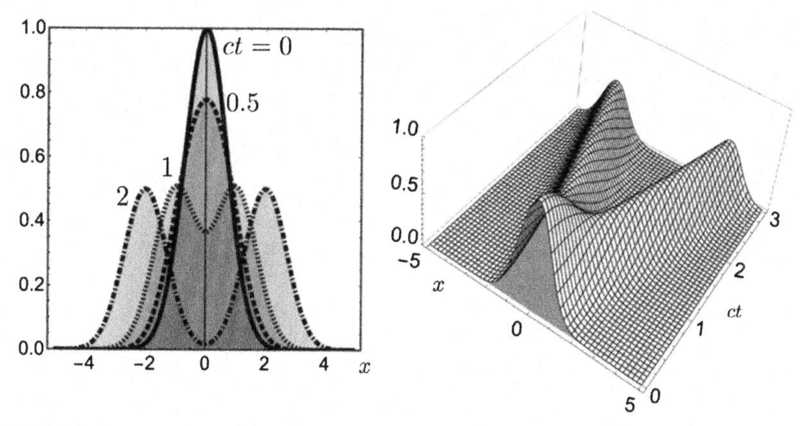

FIGURE 2.11

Example of traveling waves.

where $\omega = ck$, and k (rad/m) denotes the wave number. Similarly,

$$f(ct + x) = Ae^{ik(ct+x)} = Ae^{i(\omega t+kx)} \tag{2.39}$$

for the negative direction of x. The expression

$$f(ct \pm x) = Ae^{ik(ct\pm x)} = Ae^{i(\omega t\pm kx)} \tag{2.40}$$

is called the sinusoidal plane wave with the angular frequency of ω whose magnitude is constant independent of the space and the time variables.

Taking a traveling wave for the positive direction given by Eq. (2.38) once more, the variable $\omega t - kx$ is called the phase of the progressive wave. The phase difference θ between two points x_1 and x_2 ($x_2 > x_1$) becomes

$$\theta = (\omega t - kx_2) - (\omega t - kx_1) = -k(x_2 - x_1) < 0. \quad \text{(rad)} \tag{2.41}$$

Negative phase difference is called phase delay. The phase at x_2 is delayed by $x_2 - x_1$ from that at the position x_1.

The phase delay in proportion to the traveling distance is called the linear phase or propagation phase delay [14][15]. The linear phase delay does not change the waveforms during the propagation in the space. A plane wave travels without any change in magnitude and with linear phase. Perception of spatial distance seems unlikely by traveling plane waves in one-dimensional space.

The time required by wave propagation between a pair of spatial positions makes the phase delay or phase lag. The phase delay $k(x_2 - x_1)$ is due to the time lag $\tau = (x_2 - x_1)/c$, and the phase delay can be rewritten as

$$-k(x_2 - x_1) = -\omega\tau = -\omega\frac{x_2 - x_1}{c}. \quad \text{(rad)} \tag{2.42}$$

2.5 Wave propagation in an acoustic tube

2.5.1 Sinusoidal wave in an open tube

Transmission of vibration in a coupled oscillator is sensitive to the eigenfrequencies in the coupled components. Multiple resonances for multiple eigenfrequencies can be seen for the propagation of waves in an acoustic tube instead of a single resonance, due to mass–spring vibration. Fig. 2.12 is an image of an acoustic tube where a sinusoidal source is located at the left end, while the right end is open.

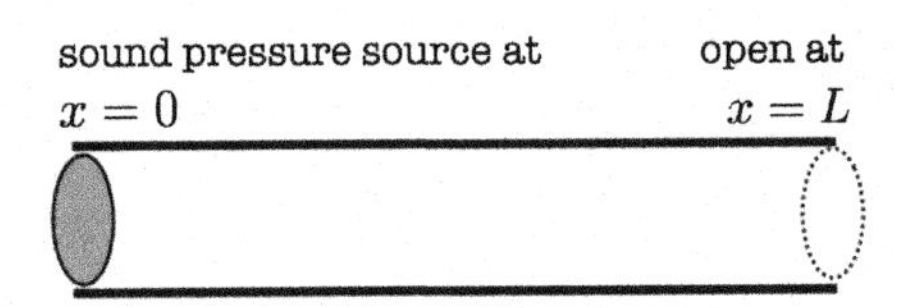

FIGURE 2.12

Image of an acoustic tube with the right end open and the left end excited by a sinusoidal pressure source.

This type of acoustic tube may be a conceptual model of a musical instrument [16].

According to the general solution of the one-dimensional wave equation given by Eq. (2.35), the traveling waves can be written as

$$p(x,t) = \left(Ae^{-ikx} + Be^{ikx}\right)e^{i\omega t} \quad \text{(Pa)} \tag{2.43}$$

in the tube. Setting the conditions at the ends (boundary conditions) such that

$$p(x,t)|_{x=0} = P_0 e^{i\omega t} \quad \text{and} \quad p(x,t)|_{x=L} = 0, \quad \text{(Pa)} \tag{2.44}$$

the solution given by Eq. (2.43) must satisfy

$$A + B = P_0 \quad \text{and} \quad Ae^{-ikL} + Be^{ikL} = 0. \quad \text{(Pa)} \tag{2.45}$$

The condition $p = 0$ at $x = L$ is called the free boundary in which the particles may freely move so that no condensation occurs and there is no sound pressure due to the sound waves [1][3][7][13][16].

Solving Eq. (2.45) for A and B,

$$p(x,t) = P_0 \frac{\sin k(L-x)}{\sin kL} e^{i\omega t} \quad \text{(Pa)} \tag{2.46}$$

is obtained, where the receiving point is located at x, while the sound source is at $x = 0$, and x denotes the distance (m) from the source. The sound pressure response at x to the sinusoidal source at $x = 0$ shows the singularities or resonances at

$$\sin k_n L = \sin \frac{\omega_n}{c} L = 0 \tag{2.47}$$

$$\omega_n = \frac{c}{2L} 2n\pi = 2\pi f_n, \quad \text{(rad/s)} \tag{2.48}$$

independent of the receiving position. The resonance is made by the periodic nature for sound wave traveling between the two ends. The fundamental frequency f_1 gives the period $T = 2L/c = 1/f_1$ (s). The resonance frequencies are composed of the fundamental and its higher harmonics. The harmonic structure in which the fundamental makes the pitch sensation explains why most musical instruments are made more or less in one-dimensional constructions.

Sound pressure propagation from the source to a receiving position, corresponding to the transmission ratio given by Eq. (2.12)

$$\frac{p(x,t)}{P_0 e^{i\omega t}} = \frac{\sin k(L-x)}{\sin kL}, \tag{2.49}$$

can be characterized by the poles (singularities) and zeros. The zeros at

$$k(L-x) = n\pi \tag{2.50}$$

depend on the distance x from the source. Fig. 2.13 illustrates the poles and zeros by using the denominator (for the poles) and the numerator (for the zeros) of Eq. (2.49), respectively.

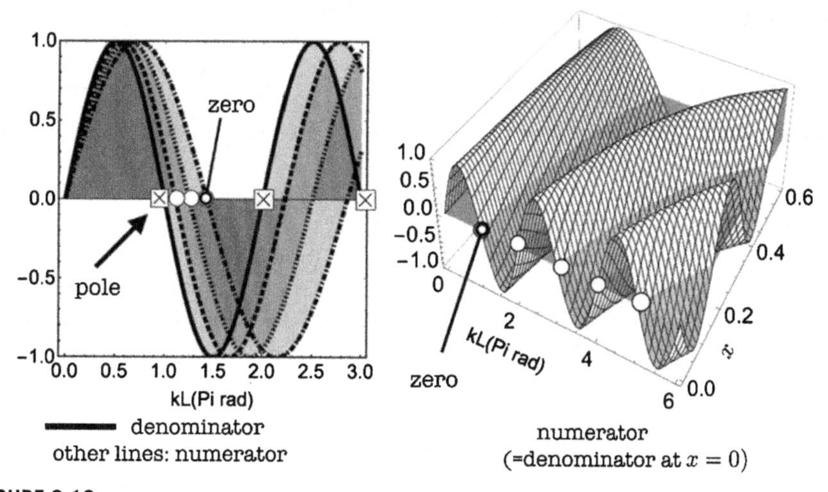

FIGURE 2.13

Illustration of the denominator and numerator of Eq. (2.49) for presenting poles and zeros.

The numerator coincides with the denominator at $x = 0$, and the poles and zeros are reduced to each other at $x = 0$. The zeros, however, move toward the high frequencies as the distance becomes longer [14][15] from the source located at the left end. In contrast the poles are independent of the receiving location.

2.5.2 Magnitude and phase responses to a sinusoidal source in an open tube

Fig. 2.14 presents the magnitude of the responses given by Eq. (2.49), but negligibly small damping is assumed, which avoids the singularities.

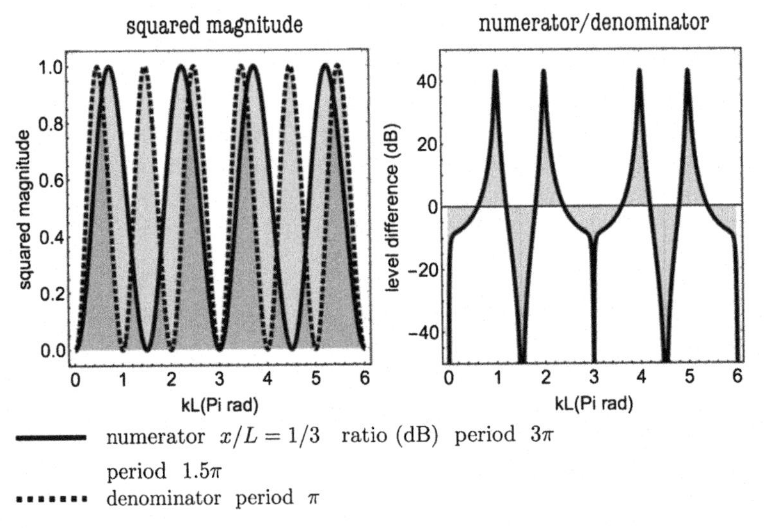

FIGURE 2.14

Magnitude characteristics for Eq. (2.49) under negligibly small damping condition.

Prominent resonant peaks can be seen around the singularities (poles), where the numerator in the ratio given by Eq. (2.49) makes zeros. Interestingly, the numerator $\sin kL(1 - x/L)$ is a periodic function of kL because of the harmonic structure of one-dimensional systems. In the example of the figure where $x/L = 1/3$ is taken, the ratio in the dB scale is periodic with the period of 3π.

On the other hand, Fig. 2.15 shows the number of zeros indicating occurrences of the sign changes of the numerator and the denominator given by Eq. (2.49). Suppose the ratio gives α times of sign changes in a frequency interval of interest. The amount of the phase accumulation becomes

$$\Phi = \pi \cdot \alpha \quad \text{(rad)} \tag{2.51}$$

because the phase shift due to a single sign change is π (rad), corresponding to an antiphase relation. The number of the sign changes in a frequency range of interest could be estimated by the difference of the numbers in the sign changes between the numerator for the zeros and the denominator for the poles. Assuming N and D for the numbers of the sign changes in the numerator and denominator (zeros-crossings), respectively, then the total number of the sign changes can be written as

$$\alpha = N - D = \frac{k(L - x)}{\pi} - \frac{kL}{\pi} = -\frac{kx}{\pi}. \tag{2.52}$$

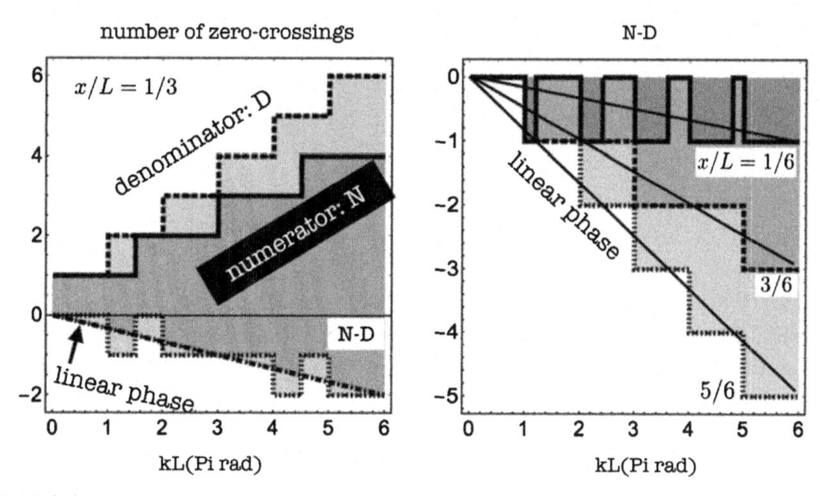

FIGURE 2.15

Number of zero-crossings in numerator and denominator of Eq. (2.49).

The total amount of phase accumulation is estimated by [14][15]

$$\Phi = \pi\alpha = -kx. \tag{2.53}$$

Interestingly, the phase trend (or smoothed phase response) is linear phase (or propagation phase), which is in proportion to k or x.

A sinusoidal wave travels like a plane wave with linearly accumulating phase trend in an acoustic tube, but with the magnitude characterized by the poles and zeros that are arranged harmonically in the frequency domain. The magnitude response makes prominent peaks around the poles or eigenfrequencies. Musical instruments are, in principle, designed according to the resonances of the poles.

2.5.3 Sinusoidal wave excited by a velocity source in an acoustic tube

In the last subsection sound waves were shown to be excited by a sound pressure source at the entrance $x = 0$ of an acoustic tube. Another type of exciting source is a velocity source.

Fig. 2.16 shows an image of an acoustic tube with an open end at $x = L$, and a sinusoidal velocity source at the entrance $x = 0$.

velocity source at $x = 0$ — open at $x = L$

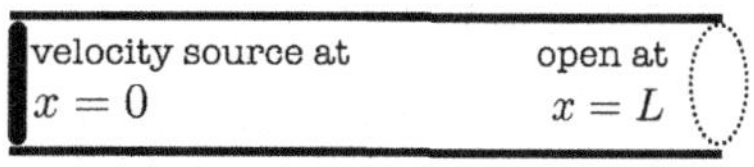

FIGURE 2.16

Acoustic tube excited by sinusoidal velocity source at $x = 0$.

The boundary condition can be formulated such that

$$p(x,t)|_{x=L}=0\ \text{(Pa)}\ \text{ and }\ \left.\frac{\partial p(x,t)}{\partial x}\right|_{x=0}=-i\omega\rho V e^{i\omega t},\ \text{(Pa/m)} \tag{2.54}$$

where

$$v=Ve^{i\omega t}\ \text{(m/s)}\ \text{ and }\ -\frac{\partial p}{\partial x}=\rho\frac{\partial v}{\partial t}=i\omega\rho v\ \text{(Pa/m)} \tag{2.55}$$

hold [1][7][13] between the sound pressure (Pa) and the particle velocity v (m/s) for sinusoidal plane waves, and ρ (kg/m^3) denotes the volume density of the medium.

Assuming the general solution given by Eq. (2.43) once more, then the solution must satisfy

$$Ae^{-ikL}+Be^{ikL}=0\ \text{(Pa)}\ \text{ and }\ Ak-Bk=\omega\rho V\ \text{(Pa/m)} \tag{2.56}$$

under the boundary condition specified by Eq. (2.54). Solving the simultaneous equation in Eq. (2.56) for A and B, then the general solution can be rewritten as

$$p(x,t)=i\rho cV\frac{\sin k(L-x)}{\cos kL}e^{i\omega t}.\quad \text{(Pa)} \tag{2.57}$$

Comparing the solution given by Eq. (2.57) with that by Eq. (2.46) for the sound pressure source, the poles of the solution for the velocity source are located at lower frequencies:

$$\cos k_n L=\cos\frac{\omega_n}{c}L=0\ \text{ or }\ \omega_n=\frac{c}{2L}(2n-1)\pi. \tag{2.58}$$

The fundamental ($n=1$) is lower than that for the pressure source by 1/1−octave. The pressure response to the velocity source is purely imaginary. In addition, the poles are composed of odd harmonics without the even harmonics. The difference in the poles between the pressure and velocity sources can be interpreted as the difference in the boundary conditions for the acoustic tube.

Fig. 2.17 displays the two types of boundary conditions for acoustic tubes: open-open and open-closed end conditions [17]. The open-open tube represents the case for the pressure source that makes the condensation or dilation of the medium at the entrance. Assuming the impulsive condensation at the left end, then the pulse-like positive pressure travels inside the tube. The pulse-like pressure wave is reflected at the right end (open end) by the negative magnitude. The return of the negative pressure wave to the left end changes the sign of the pressure to positive, which goes toward the right end. The periodic traveling yields the fundamental frequency.

In contrast, the propagation wave from a velocity source can be interpreted as the traveling waves in an acoustic tube under the open-end boundary conditions, as shown in the right panel of Fig. 2.17. Velocity excitation might be made by piston motion of a "plate" at one end of the tube. As happens in the open-open tube, the reflected pressure wave with the negative sign comes back to the source end. However,

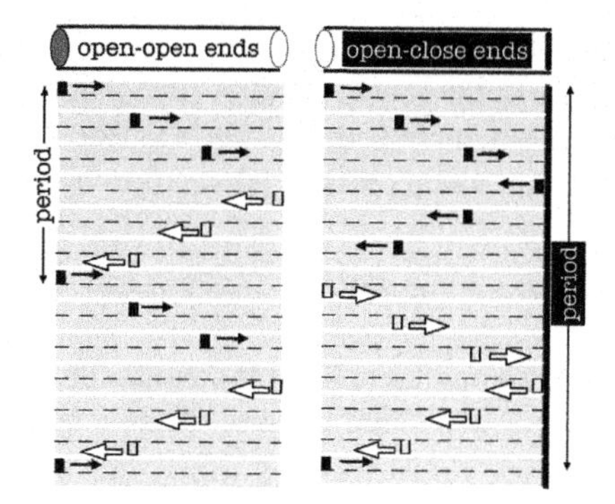

FIGURE 2.17

Acoustic tubes with open-open and open-closed ends, where pulse-like waves propagate between the two ends; from Figs. 4.1 and 4.5 [17].

no sign change occurs at the end because the end is closed by the piston plate, which reflects the wave without a sign change. This traveling of the pulse-like wave is periodic; however, the period is two times longer than that for the open-open tube. The longer period makes the fundamental lower by 1/1– octave than that for open-open tube [17].

The two types of boundary conditions may represent the conceptual models for a flute by the open-open condition, and a clarinet by the closed-open condition. The difference of the boundary conditions might explain the difference in the fundamental frequencies with their harmonics. However, the boundary condition for a clarinet might be mixed rather than purely open-closed [5]. The harmonics can be composed of even and odd harmonics even for a clarinet.

2.5.4 Poles and zeros of driving-point impedance

The ratio of sound pressure p (Pa) to volume velocity q (m^3/s) at a receiving position defines acoustic impedance ($\mathrm{Pa}\cdot\mathrm{s}/\mathrm{m}^3$). Assuming the time dependence of $\mathrm{e}^{\mathrm{i}\omega t}$, the volume velocity $q = vS$ (m^3/s) is written as

$$\mathrm{i}\omega\rho q = -\frac{\partial p}{\partial x}S = kP_0S\frac{\cos k(L-x)}{\sin kL}\cdot \mathrm{e}^{\mathrm{i}\omega t} \quad (\mathrm{Pa}\cdot\mathrm{m}) \tag{2.59}$$

in the acoustic tube excited by a sound pressure source at $x = 0$, where S (m^2) gives the area of the tube. The acoustic impedance for the open tube is given by

$$Z_{A_P} = \frac{p}{q} = \frac{\mathrm{i}\rho c}{S}\frac{\sin k(L-x)}{\cos k(L-x)}, \quad (\mathrm{Pa}\cdot\mathrm{s}/\mathrm{m}^3) \tag{2.60}$$

which depends on the position in the tube. The acoustic impedance becomes

$$Z_{0_{A_P}} = \frac{\rho c}{S} \quad (\mathrm{Pa}\cdot\mathrm{s}/\mathrm{m}^3) \tag{2.61}$$

for a plane wave in a free field that is independent of the observation position.

The acoustic impedance at the source position ($x = 0$ in the example of the tube) is called the acoustic driving-point impedance. The acoustic driving-point impedance for the open tube excited by a sound pressure source becomes

$$Z_{dA_P} = \frac{i\rho c}{S} \tan kL, \quad (\text{Pa} \cdot \text{s/m}^3) \tag{2.62}$$

where the poles and zeros are interlaced [14][15]. No accumulation of the phase can be observed as the frequency goes high because of the interlacing of the poles and zeros.

The acoustic driving-point impedance can be defined for the velocity-source excitation, too. Taking the ratio of the sound pressure to the volume velocity at the source position,

$$Z_{dA_V} = \frac{i\rho c}{S} \tan kL = Z_{dA_P} = Z_{dA} \quad (\text{Pa} \cdot \text{s/m}^3) \tag{2.63}$$

is obtained. Interestingly, the acoustic driving-point impedance for the velocity source is same as that for the sound pressure source. Again, the poles are interlaced with the zeros. Recalling that

$$p = Z_{dA} \cdot q \ (\text{Pa}), \quad \text{or} \quad q = \frac{p}{Z_{dA}}, \quad (\text{m}^3/\text{s}) \tag{2.64}$$

the resonance frequencies, however, are different under the conditions between pressure and velocity source excitation. The poles of the acoustic driving-point impedance make the resonance for the velocity source excitation, while the zeros render the resonance for the pressure source excitation.

Sound wave propagation in a medium is a process that requires energy to work on exciting adjacent parts of the medium. The energy is supplied by a sound source. The idealized acoustic tube has an idealized end at which the sound wave is perfectly reflected without energy consumption. No energy consumption indicates that no sound is radiated from such an idealized boundary. Assuming the sound is traveling in an ideal tube, the traveling continues periodically without decaying or radiation even after the sound source has stopped.

The purely imaginary function given by Eq. (2.63) for the acoustic driving-point impedance is a result of the idealized boundary condition. The sound pressure and velocity are out-of-phase at the sound source position. Recalling the orthogonality as

$$\frac{1}{T} \int_0^T \sin \omega t \cos \omega t dt = 0, \tag{2.65}$$

where T (s) denotes the period, then no energy is consumed or no sound is radiated in the out-of-phase condition. The ratio of the sound pressure to the velocity is always real for a plane wave, as shown in Eq. (2.61). The sound propagates in the medium as long as the energy is supplied by a sound source.

In addition to the acoustic impedance, the mechanical impedance is defined by the ratio

$$Z_M = \frac{f}{v}, \quad (\mathrm{N \cdot s/m}) \tag{2.66}$$

where f (N) and v (m/s) denote the force and particle velocity, respectively. The acoustic or mechanical impedance is used case by case, as mentioned in Section 1.4.3 [1]. The specific impedance is also defined by the ratio of sound pressure to particle velocity,

$$Z_0 = \frac{p}{v} = \rho c, \quad (\mathrm{Pa \cdot s/m}) \tag{2.67}$$

for a plane wave traveling in a medium, where ρ (kg/m^3) denotes the volume density of the medium and c (m/s) gives the speed of sound in the medium. The specific impedance is used as a representative of the medium in which a plane wave travels.

2.6 Exercises

1. Derive Eq. (2.1).

2. Show that Eq. (2.3) satisfies Eq. (2.2).

3. Derive Eq. (2.12).

4. Confirm that Eq. (2.20) is a solution of Eq. (2.18).

5. Compare the resonance frequency given by Eq. (2.24) with the frequency of the decaying oscillation ω_d given by Eq. (2.21).

6. Derive Eq. (2.29).

7. Confirm that Eq. (2.35) is a general solution of a one-dimensional wave equation.

8. Derive Eqs. (2.46), (2.57), (2.62), and (2.63).

9. Show the acoustic driving-point impedance for an open-open tube, for example [1][13][14][15]

$$Z_{dA} = \frac{p(x',t)}{q(t)} = \frac{i\rho c}{S} \cdot \frac{\sin kx' \cdot \sin k(L-x')}{\sin kL}, \tag{2.68}$$

where $x'(\neq 0)$ denotes the position of the velocity source, L is the length of the tube, S gives the cross-sectional area of the tube, ρ shows the density of the medium in the tube, c is the speed of sound in the tube, and q gives the volume velocity of the velocity source.

10. Show that no phase accumulation is yielded in the acoustic driving-point impedance above [1][14][15].

11. Explain the terminologies listed below.
(1) single-degree-of-freedom system (2) frequency of free vibration (3) resonance frequency (4) eigenfrequency (5) antiphase (6) in phase (7) complex frequency plane (8) transfer function (9) poles and zeros of transfer function (10) phase accumulation by poles and zeros (11) impulse response and convolution (12) transient and steady-state responses (13) wave equation in one-dimensional space (14) propagation phase (15) fundamental and harmonics (16) acoustic or mechanical impedance (17) driving-point impedance (18) specific impedance (19) out-of-phase

References

[1] M. Tohyama, Sound and Signals, Springer, 2011.
[2] Y. Hirata, T. Kawai, On the isolation of solid-borne noise from a water pump (in Japanese), Report of Architectural Acoustics Research Meeting AA78-20, Acoust. Soc. Japan, 1978.
[3] T.D. Rossing, N.H. Fletcher, Principles of Vibration and Sound, Springer, Heidelberg, 1995.
[4] Y. Hara, Phase relation of loudspeaker system, private communication.
[5] M. Tohyama, Sound in the Time Domain, Springer, 2017.
[6] M. Tohyama, R.H. Lyon, Zeros of a transfer function in a multi-degree-of-freedom system, J. Acoust. Soc. Am. 86 (5) (1989) 1854–1863.
[7] J. Blauert, N. Xiang, Acoustics for Engineers, Springer, 2008.
[8] M.R. Schroeder, Models of hearing, Proc. IEEE 63 (9) (September 1975) 1332–1350.
[9] B.C.J. Moore, An Introduction to the Psychology of Hearing, Academic Press, 1997.
[10] S.J. Elliott, C.A. Shera, The cochlea as a smart structure, Smart Mater. Struct. 21 (2012) 064001.
[11] Y. Hirata, The frequency discrimination of the ear, Relevant Articles, wavesciencestudy.com, Nov. 06 (2014).
[12] T. Mori, Gendai no koten kaiseki (Modern-classical mathematics) (in Japanese), Chikuma Shobo, 2006.
[13] M. Tohyama, T. Koike, Fundamentals of Acoustic Signal Processing, Academic Press, 1998.
[14] R.H. Lyon, Progressive phase trends in multi-degree-of-freedom systems, J. Acoust. Soc. Am. 73 (4) (1983) 1223–1228.
[15] R.H. Lyon, Range and frequency dependence of transfer function phase, J. Acoust. Soc. Am. 76 (5) (1984) 1433–1437.
[16] T.D. Rossing, N.H. Fletcher, The Physics of Musical Instruments, Springer, Heidelberg, 1988.
[17] C. Taylor, Exploring Music, IOP Publishing Ltd., 1992.

CHAPTER

Modulation waveform and masking effect

3

CONTENTS

3.1 Analytic signals and envelopes

3.1.1 Analytic representation of sinusoidal function

Take a sinusoidal function such as

$$y(t) = \sin \omega t \tag{3.1}$$

in a real function, where $\omega > 0$ gives the angular frequency (rad/s). The function can be expressed in a complex function as

$$y(t) = \frac{e^{i\omega t} - e^{-i\omega t}}{2i} \tag{3.2}$$

where $\pm\omega$ is called a positive or negative angular frequency. The exponential function $e^{i\omega t}$ $(\omega > 0)$ shows a unit circle centered at the origin with the radius of unity in the complex plane. The positive angular frequency indicates the phase angle proceeds counterclockwise, while the function $e^{-i\omega t}$ denotes the clockwise unit circle with the negative phase angle.

Acoustic Signals and Hearing. https://doi.org/10.1016/B978-0-12-816391-7.00011-5

Taking the other sinusoidal function

$$x(t) = \cos \omega t, \tag{3.3}$$

then

$$\cos \omega t = \frac{\mathrm{e}^{\mathrm{i}\omega t} + \mathrm{e}^{-\mathrm{i}\omega t}}{2} \tag{3.4}$$

holds as a superposition of counterclockwise and clockwise unit circles again. The complex functions $z_c(t)$ and $z_s(t)$ such that

$$\cos \omega t = \Re[z_c(t)] = \Re[\mathrm{e}^{\mathrm{i}\omega t}] = \Re[\cos \omega t + \mathrm{i} \sin \omega t] \tag{3.5}$$

$$\sin \omega t = \Re[z_s(t)] = \Re[-\mathrm{i}\mathrm{e}^{\mathrm{i}\omega t}] = \Re[\sin \omega t - \mathrm{i} \cos \omega t] \tag{3.6}$$

are called analytic signal representation of sinusoidal functions, respectively.

Analytic signals $z_c(t), z_s(t)$ are complex functions; however, those are composed of only the positive angular frequencies. The analytic signals are formally derived by discarding the negative frequency components from the complex functions. Taking only the positive frequency component from Eq. (3.4), then

$$\hat{z}_c(t) = \frac{1}{2}\mathrm{e}^{\mathrm{i}\omega t} = \frac{1}{2} z_c(t) \tag{3.7}$$

is obtained. Similarly,

$$z_s(t) = 2\hat{z}_s(t) \tag{3.8}$$

is derived, where

$$\hat{z}_s(t) = \frac{1}{2\mathrm{i}}\mathrm{e}^{\mathrm{i}\omega t} = \frac{-\mathrm{i}}{2}\mathrm{e}^{\mathrm{i}\omega t}. \tag{3.9}$$

Taking the magnitudes (or absolutes)

$$|z_c(t)| = \left|\mathrm{e}^{\mathrm{i}\omega t}\right| = 1 \quad \text{and} \quad |z_s(t)| = \left|-\mathrm{i}\mathrm{e}^{\mathrm{i}\omega t}\right| = 1. \tag{3.10}$$

Magnitudes of analytic signals are called the envelopes or Hilbert envelopes of waveforms. Main concern of introducing the analytic representation to a waveform would be mathematically defining the envelope for the waveform. The results given by Eq. (3.10) indicate flat envelopes for sinusoidal waveforms independent of the time variable. The flat envelopes would be acceptable for hearing sinusoidal signals, where no temporal variations in the magnitude are sensed for a single sinusoidal wave independent of the frequency.

The analytic signal representation newly creates the imaginary part that is added to the original real waveform. The imaginary part could be interpreted as a $-\pi/2$–phase shift. Taking a sinusoidal function $x(t) = \cos \omega t$, then the imaginary part could be derived as

$$y(t) = \cos(\omega t - \pi/2) = \sin \omega t. \tag{3.11}$$

The analytic representation of $x(t)$ could be derived as

$$z_c(t) = x(t) + \mathrm{i}y(t) = \mathrm{e}^{\mathrm{i}\omega t}. \tag{3.12}$$

Similarly, for $x(t) = \sin \omega t$

$$y(t) = \sin(\omega t - \pi/2) = -\cos \omega t \tag{3.13}$$

$$z_s(t) = x(t) + \mathrm{i}y(t) = -\mathrm{i}\mathrm{e}^{\mathrm{i}\omega t}. \tag{3.14}$$

The definition of envelopes by using analytic representation of signals could be extended to compound signals that are composed of sinusoidal waves. Take an example such that

$$x(t) = \cos \omega t + \sin 2\omega t = x_1(t) + x_2(t). \tag{3.15}$$

The analytic formula for $x(t)$ becomes

$$z(t) = \mathrm{e}^{\mathrm{i}\omega t} - \mathrm{i}\mathrm{e}^{\mathrm{i}2\omega t} = \mathrm{e}^{\mathrm{i}\omega t}(1 - \mathrm{i}\mathrm{e}^{\mathrm{i}\omega t}). \tag{3.16}$$

Taking the squared magnitude of $z(t)$ (or squared Hilbert envelope)

$$|z(t)|^2 = |1 - \mathrm{i}\mathrm{e}^{\mathrm{i}\omega t}|^2 = 2(1 + \sin \omega t). \tag{3.17}$$

Fig. 3.1 shows an example of the Hilbert envelope described by Eq. (3.17).

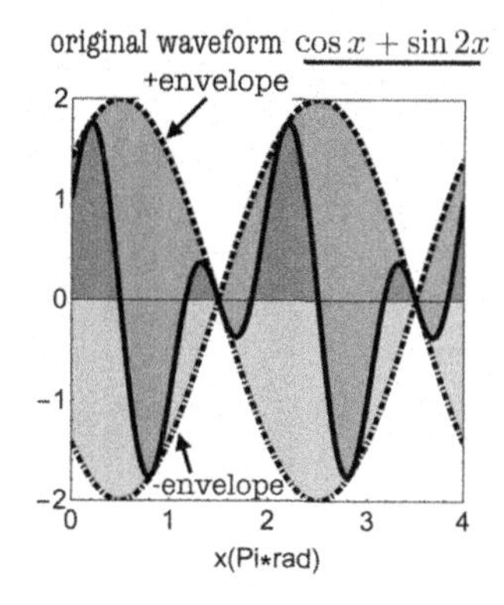

FIGURE 3.1

Example of the Hilbert envelope of a compound sinusoidal signal where ωt is replaced by x.

Comparing the Hilbert envelope with the original signal $x(t)$ the intension of the Hilbert envelope can be intuitively understood.

3.1.2 Modulation waveform by analytic signal representation

The analytic signal representation can be extended into other waveforms. A real signal can be assumed as a superposition of the positive and negative frequency components, subject to the time average of the signal being removed. Discarding the negative frequency components, and taking the positive frequency components twice, then the analytic signal can be derived as

$$z(t) = x(t) + iy(t) = |z(t)|e^{i\theta(t)}, \tag{3.18}$$

where $z(t)$ is composed of only the positive frequency components, $x(t)$ denotes the original real function, and $y(t)$ is the newly created imaginary part of the analytic signal. The real part of the analytic signal, $x(t)$, can be expressed as

$$x(t) = |z(t)| \cos\theta(t), \tag{3.19}$$

where $|z(t)|$ is called the instantaneous magnitude of the analytic signal, $\cos\theta(t)$ is called the carrier, and $\theta(t)$ denotes the instantaneous phase of the analytic signal. The representation of a signal $x(t)$ given by Eq. (3.19) is called a modulation form.

Fig. 3.2 shows an example of the modulation form of a waveform given by Eq. (3.19) or the analytic representation of a signal.

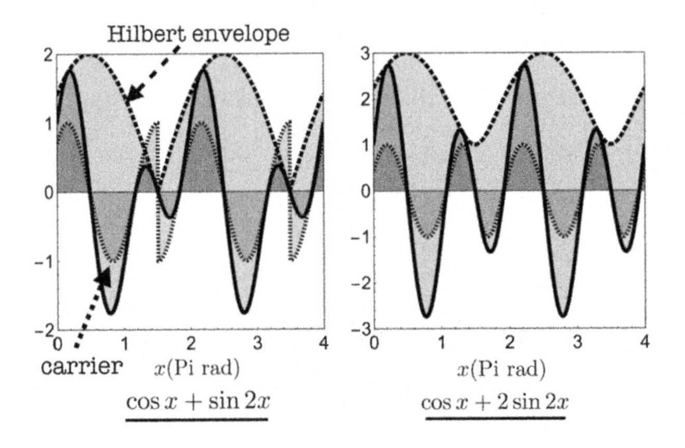

FIGURE 3.2

Example of modulation form of a real signal where ωt is replaced by x.

The envelope or the instantaneous magnitude shows the temporal variations in the signal magnitude, and indicates the signal dynamics in the time domain [1]. The carrier expresses the fine structure of the signal represented by zero-crossings or the temporal fluctuations of the frequency in the time domain, where $d\theta/dt$ gives the instantaneous angular frequency. The left panel in Fig. 3.2 shows the carrier that could be discontinuous when the Hilbert envelope becomes zero. The instantaneous phase makes the phase jump due to the zero of the envelope. The jump of the instantaneous phase in the time domain could be understood similarly to phase accumulation in the frequency domain by a zero [2].

On the other hand, the right panel shows the continuous carrier when the Hilbert envelope is positive (no zeros) such that

$$|z(t)|^2 = 5 + 4\sin\omega t > 0, \tag{3.20}$$

where

$$x(t) = \cos\omega t + 2\sin 2\omega t. \tag{3.21}$$

3.1.3 Amplitude modulation

Amplitude modulation is formally defined as

$$x(t) = (1 + \cos\omega_e t)\cos\omega_c t = |z(t)|\cos\theta(t), \tag{3.22}$$

where

$$|z(t)| = |A(t)| = A(t) = 1 + \cos\omega_e t \tag{3.23}$$

and $A(t)$ is nonnegative such that

$$A(t) = 1 + \cos\omega_e t \geq 0. \tag{3.24}$$

The amplitude modulation expressed by Eq. (3.22) is composed of three sinusoidal components such that ω_c and $\omega_c \pm \omega_e$. This makes the difference in the zeros of the envelopes between Eqs. (3.17) and (3.23). Actually, the envelope given by Eq. (3.17)

$$|z(t)|^2 = 2(1 + \sin\omega t) = B^2(t) = \left(\pm\sqrt{2(1 + \sin\omega t)}\right)^2 \tag{3.25}$$

makes the difference from $A(t)$ such that $B(t)$ could be negative, and that the envelope $|B(t)|$ is not a smooth function. The right panel of Fig. 3.3 illustrates an example of an amplitude-modulation envelope with its smooth carrier function even when the envelope becomes zero. The instantaneous phase does not make the phase jump different from the case in Eq. (3.17).

3.1.4 Envelope and instantaneous phase

Suppose the analytic representation of $x(t)$ to be $z(t) = |z(t)|e^{i\theta(t)}$. An interesting question would be whether or not the instantaneous phase $\theta(t)$ could be derived from $|z(t)|$ [2]. The imaginary part of the analytic representation could be obtained from the real part, $x(t)$, by discarding the negative frequency components. A signal $s(t)$ defined as

$$s(t) = \begin{cases} s(t) & t \geq 0 \\ 0 & t < 0 \end{cases} \tag{3.26}$$

is called a causal signal. The spectral function of the causal signal $s(t)$, or Fourier transform of the signal, is expressed as

$$S(\omega) = S_r(\omega) + \mathrm{i}S_r(\omega) = |S(\omega)|\mathrm{e}^{\mathrm{i}\phi(\omega)}, \tag{3.27}$$

where $S_r(\omega)$ and $S_i(\omega)$ give the real and imaginary parts of $S(\omega)$, respectively. According to the property of the Fourier transform of a causal signal, the real and imaginary parts cannot be independent of each other.

The relationship between the magnitude $|S(\omega)|$ and the phase $\phi(\omega)$ of the Fourier transform is a different story, however. If the causal signal is of minimum phase, then the magnitude and phase could be dependent on each other (see section 6.4) [2][3][4]. The similar relationship could be read between the instantaneous envelope and instantaneous phase in the complex time domain. The real and imaginary parts of an analytic signal could be dependent on each other because the spectral function has no negative frequency components, namely causal in the frequency domain. If the analytic signal is of minimum instantaneous phase in the complex time domain, then the instantaneous envelope (or magnitude) cannot be independent of each other [2][5].

3.1.5 Effects of noise on envelope

Envelopes that represent the signal dynamics in the time domain are closely related to the perception of speech, known as speech intelligibility [6]. The dynamical behavior of signals is lost as the envelopes are distorted. The right panel of Fig. 3.3 illustrates an image of the deterioration of the amplitude modulation envelopes by the background noise.

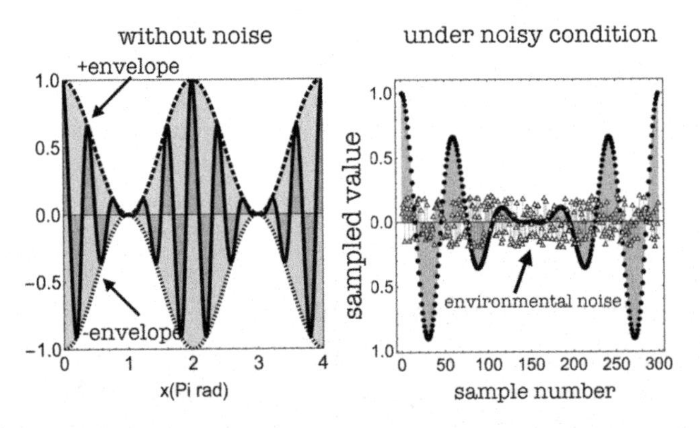

FIGURE 3.3

Image of modulation envelopes under noisy conditions.

Speech intelligibility under noisy conditions can be estimated in terms of the modulation index [6].

3.2 Speech intelligibility under masked condition

3.2.1 Intelligibility of noisy speech

A noisy speech signal is called masked speech by a masker or random noise. One of the maskers would be a wideband random noise such as white noise. The modulation index (defined in Eq. (3.28)) decreases as the energy of the noise increases, as intuitively understood from Fig. 3.3 in the last section. However, the masking effect of the white noise masker is not prominent. Fig. 3.4 is an example of intelligibility under masked conditions, where TMR stands for the target-to-masker energy ratio [7].

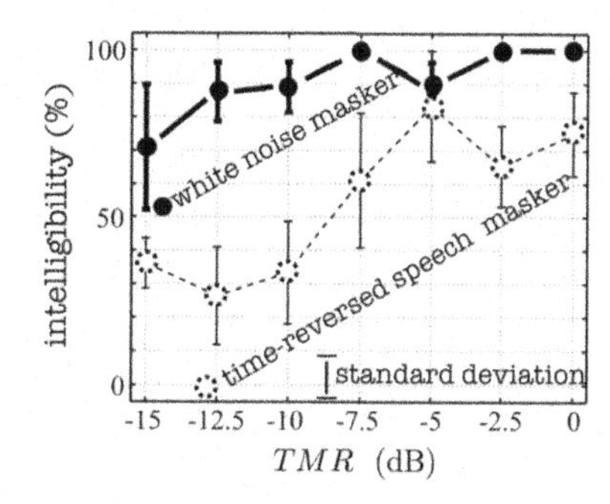

FIGURE 3.4

Intelligibility scores for words under masked conditions by white noise (filled circle) and time-reversed speech (open circle) from Fig. 2 [7].

The experimental conditions are summarized in Fig. 3.5.

1 (s)	2 (s)	1 (s)
silence	**target under masker**	silence

listening condition	diotic
maskers	white noise or reverse speech
subjects	9
TMR (dB)	0, -2.5, -5, -7.5, -10, -12.5 -15
test signals	35 (for each masker)

FIGURE 3.5

Experimental conditions for Fig. 3.4 from Fig. 1 [7].

The subjects who participated in the experiments are native Japanese speakers who listened to the test samples under diotic condition (i.e., no difference in the signal to the ears, through headphones). The subjects were asked to write down Japanese phrases composed of two words (a noun and an adjective) as they listened to mixtures of the target and the masker. The details of the conditions are described in the reference [7].

Modulation indexes for the envelopes can be lower than unity [6] if the noise maskers are superposed. However, the loss of intelligibility is not remarkable, as shown by the filled circles in Fig. 3.4, even if the masker energy becomes higher. Suppose a modulation envelope of

$$e(t) = \frac{1}{2}(1 + m\cos\omega_e t), \tag{3.28}$$

where $m(0 < m \leq 1)$ denotes the modulation index. The waveform of the envelope might always be preserved after discarding its average, even if the modulation index decreases as the superposed masker's energy increases. The preservation of the waveform of the envelope implies that the masking effects on the intelligibility are not prominent.

3.2.2 Effects of information masking on intelligibility

The intelligibility of noisy speech decreases as the TMR becomes lower. A masker that causes its masking effect in proportion to TMR is called an energy masker. A representative example of the energy masker would be white noise, although the masking effect on speech intelligibility might not be remarkable. Another type of masker, whose effect is almost independent of TMR, is possible in principle and is called an informational masker [7][8][9][10], that may cause the masking effect almost independent of TMR, in principle.

Time-reversed speech is an example of an informational masker [8]. Suppose a real signal $x(t)$. Its time-reversed signal is expressed as $x(-t)$. A mixture of the original and time-reversed signals becomes

$$y(t) = x(t) + x(-t), \tag{3.29}$$

which is an even (or symmetric) function of the time variable t. The change into an even envelope from the original envelope may dramatically reduce the intelligibility.

The open circles in Fig. 3.4 shows the intelligibility scores for the mixtures of speech and time-reversed signals [7]. The time-reversed maskers, however, are not made by the target phrases, but different maskers are created by the same talker because the time-reversed signal of the target seems too idealized as a masker from the frame of reference of daily speech communication. The experimental conditions are the same as those for the white noise masker, as shown in Fig. 3.5, except that the subjects were asked to write down only one phrase (composed of a noun and an adjective) because the subjects might be able to hear two different words for mixtures of the original and time-reversed speech.

In spite of the approximated time-reversed maskers, the masking effect on the intelligibility is much more prominent than that caused by the white noise masker.

The decrease in intelligibility caused by the time-reversed speech masker implies that the envelopes of masked speech may largely change from the original envelopes. The similarity of a pair of random variables X and Y can be estimated by the cross-correlation coefficient between the pair of variables. The cross-correlation coefficient is defined as

$$R_{XY} = \frac{\mathrm{E}[X \cdot Y]}{\sqrt{\mathrm{E}[X^2]}\sqrt{\mathrm{E}[Y^2]}}, \tag{3.30}$$

where E[$*$] denotes taking the ensemble average of the variable ($*$), and

$$\mathrm{E}[X] = \mathrm{E}[Y] = 0. \tag{3.31}$$

Recalling the model of the hearing organ as a set of band-pass filters [11], the effect of the maskers on the intelligibility can be estimated on a frequency-band basis as 1/4−octave band width (25 bands from 0.125 − 8 (kHz)).

Fig. 3.6 shows the cross-correlation coefficients of the squared envelopes, as a function of TMR, between the speech materials with and without time-reversed maskers, where the correlation coefficients are averaged from 0.125 − 8 (kHz) for five speech utterances under every condition [7].

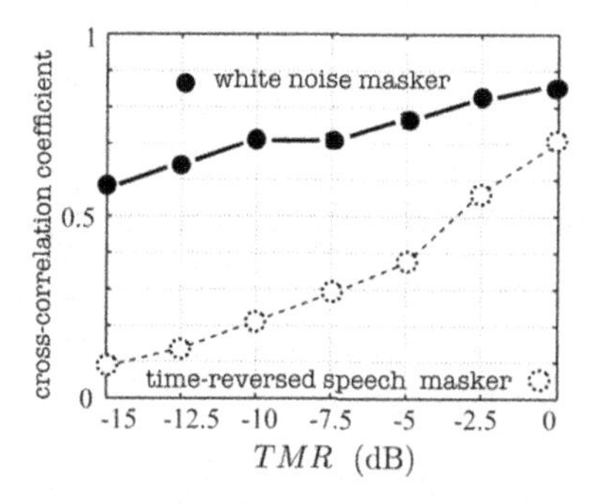

FIGURE 3.6

Averaged cross-correlation coefficients between squared narrowband envelopes of target and masked target from Fig. 3 [7].

The correlations for the squared envelopes are shown as filled and open circles. The correlations for the white noise masker given by the filled circles are in the range of 0.6 − 0.8, and imply that the envelope can be preserved even under the masked condition. In contrast, the correlations shown by the open circles rapidly decrease from 0.7 to 0.1 for the time-reversed masker as TMR decreases. The envelope of the target is largely changed by the time-reversed masker.

The preservation of the narrowband envelopes (such as 1/4−octave band envelopes) in addition to the modulation index keeps the speech intelligible. The energy maskers reduce the modulation index, but keep the modulation waveform. In contrast, informational maskers greatly change the envelopes.

Fig. 3.7 suggests that the intelligibility changes as a function of the envelope cross-correlation coefficients.

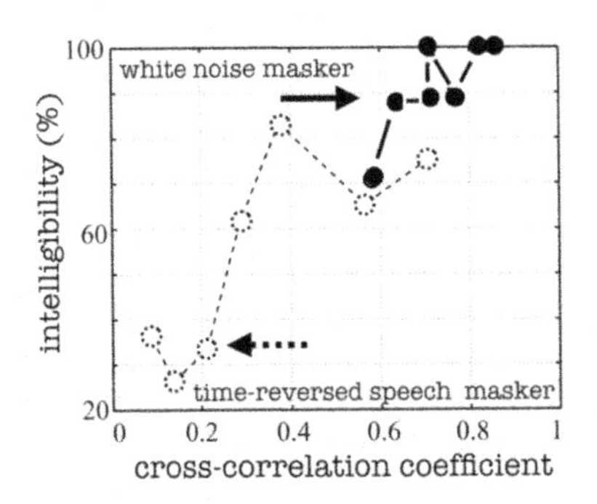

FIGURE 3.7

Intelligibility and cross-correlation coefficients for squared envelopes under masked conditions by white noise or time-reversed speech masker from Fig. 4 [7].

High intelligibility scores (filled circles) of the energy masker (white noise) are distributed in the narrower range of higher correlations, while a wide range of intelligibility scores (open circles) of the informational masker (time-reversed speech) can be seen under the broader range of lower correlations.

3.3 Spectral fine structure of masked speech

3.3.1 Carriers for intelligible speech

The modulation indexes and narrowband envelopes convey intelligibility of speech. On the other hand, the narrowband carriers are mostly responsible for the tonal character of speech waveforms [12]. Substituting the 1/4−octave band random noise for the corresponding carrier at every frequency band, then the tonal character of the speech is almost lost, although the synthesized speech is still intelligible. Replacing the carrier by the sinusoidal wave of the center frequency at every 1/4−octave band, then the tonality is almost recovered; however, it still does not sound natural.

The results of changing the carriers imply the periodic or harmonic nature of speech waveforms are conveyed by the carrier. The periodic characteristics of a narrowband noise such as 1/4−octave band noise may be too weak or fluctuating, while the sinusoidal carrier of the center frequency at every frequency band is too steady without any fluctuations in the carrier [12].

Frame-wise sinusoidal carriers can be taken by the maximum power spectral components at every frame in each frequency band. The detailed scheme for the frame-wise carrier selection can be seen in reference [13]. The frame-wise carriers would be helpful to natural sounding speech that is almost perfectly intelligible. The spectral dynamics would be interesting from the frame of reference of masking effects in addition to the narrowband envelopes in the time domain [7][14].

3.3.2 Frame-wise auto-correlation sequences in spectral domain

The sinusoidal component with the maximum power spectrum can be taken at every short frame. Extending the number of prominent spectral components, the frame-wise harmonic structure can be estimated by the spectral auto-correlation sequence. Frame-wise auto-correlation analysis in the frequency domain would be interesting in order to represent masking effects on speech waveforms by energy or informational maskers [7][14]. Fig. 3.8 shows a schematic of the frame-wise spectral auto-correlation analysis for the target or masked speech waveforms [7].

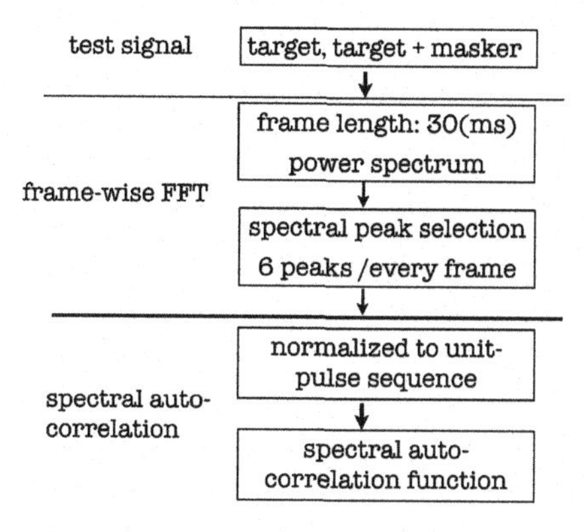

FIGURE 3.8

Scheme of frame-wise spectral auto-correlation analysis.

The major power spectral components are normalized with unit-power spectra, as shown in Fig. 3.9, so that unit-pulse-like sequences are created on the frequency axis frame by frame.

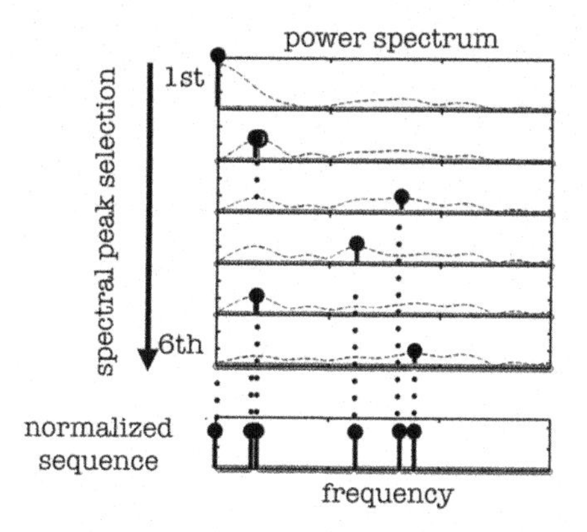

FIGURE 3.9

Example of a unit-pulse sequence representing spectral spacing of major spectral components.

The auto-correlation sequence of $x(n)$ can be written as

$$r(n) = \sum_{m} x(m)x(m-n), \tag{3.32}$$

which gives the histogram of inter-pulse spacings if $x(n)$ is a unit-pulse sequence. The most probable spacing gives an estimate of the fundamental period of $x(n)$ [14].

Fig. 3.10 displays the spectral auto-correlation sequences for the target and masked target by the white noise or time-reversed speech masker.

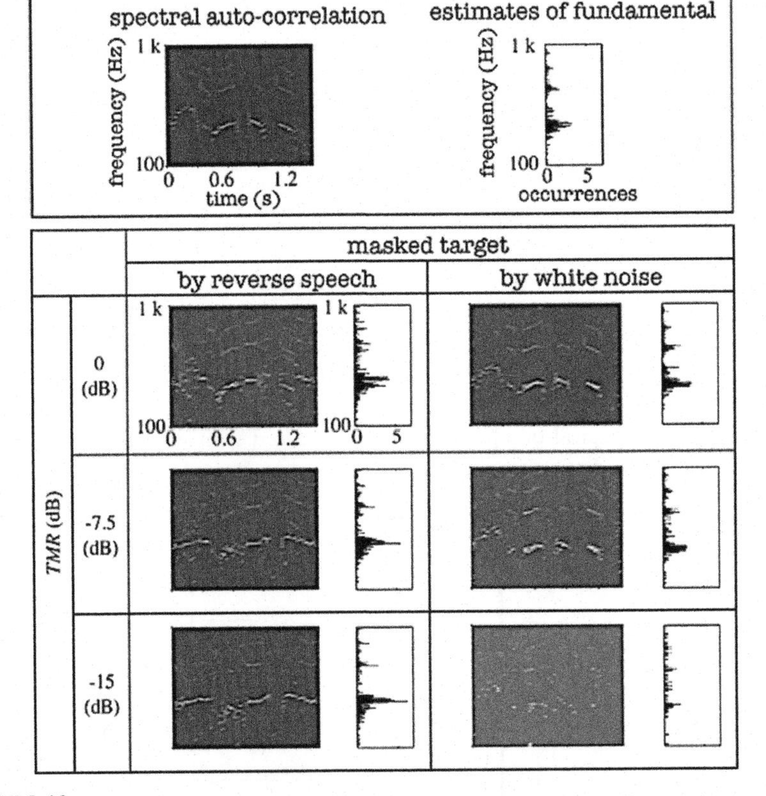

FIGURE 3.10

Spectral auto-correlation sequences of target and masked target by white noise or time-reversed speech masker from Fig. 6 [7].

The fundamental and its harmonics can be seen for the target in the upper panel as a function of the frame position (or the time). The harmonic structure can be confirmed in the histogram of the estimates of fundamental frequencies by the peaks of spectral correlation sequences frame by frame.

In contrast, the frame-wise spectral auto-correlation sequences are presented for the masked target in the lower panel, where the left column shows the results from the time-reversed masker, and the right column gives them for the white noise masker. Interestingly, the histograms of the estimates of the fundamental frequencies are mostly similar to that for the target. However, the temporal change of the fundamental frequencies with the harmonics may not be clearly traced under the time-reversed masker. The speech material under the time-reversed masker might still be perceived as speech-like sound; however, the material quite likely will not be intelligible. Temporally changing spectral structures may be significant in the preservation of intelligible speech and in the preservation of narrowband envelopes.

3.3.3 Deformation of harmonic structure

The frame-wise spectral auto-correlation sequence may represent the time-variant harmonic structure. Dissimilarity between a pair of spectral auto-correlation sequences under the unmasked (target without masker) and masked conditions might express the masking effects on the harmonic structure.

Taking the frame-by-frame cross-correlation coefficients between the spectral auto-correlation sequences, a random sequence of the cross-correlation coefficients can be obtained under the masked conditions [7]. The left panel of Fig. 3.11 shows the cumulative distributions of the frame-wise cross-correlation coefficients for the masked target created by the white noise (filled dots) and time-reversed speech (open dots) maskers, respectively. Low correlation coefficients are more likely for the time-reversed masker under most of the TMR conditions than for the white noise masker. The cross-correlation coefficients are mostly distributed under 0.3 for the time-reversed masker, while the coefficients almost uniformly distribute in the correlation range for the white noise masker. The right panel similarly displays the $75th$ percentile scores of the frame-wise cross-correlation coefficients. Both panels show the differences in the masking effects between the maskers from the perspective of deformation in the harmonic structures.

Fig. 3.12 illustrates the intelligibility scores for the masked speech samples as a function of the $75th$ percentile scores given in the right panel of Fig. 3.11. The $75th$ percentile distributes in a wide range $(0.2 - 0.8)$ of the cross-correlation coefficients. The white noise masker reveals the distribution in the higher correlation range, and masked speech samples could be almost intelligible even under the masked conditions. In contrast, the time-reversed masker makes the target mostly unintelligible by yielding much lower correlation coefficients with respect to the harmonic structure. Figs. 3.6 (preservation of narrowband envelopes) and 3.12 (harmonic structure) both show the difference between the effects of energy and informational masking on intelligibility.

Narrowband envelopes can be extended into the time-variant power spectral behavior as a limit when the bandwidth becomes narrow enough to represent time-dependent line spectral components. On the other hand, the spectral harmonic structure, represented by the auto-correlation when the major spectral components are

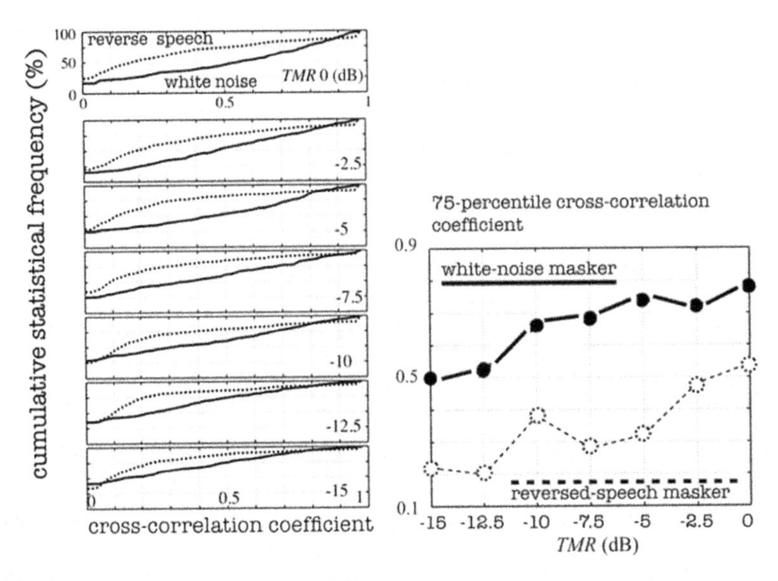

FIGURE 3.11

Cumulative distributions (left panel) and 75*th* percentile (right) for frame-wise cross-correlation coefficients between spectral auto-correlation sequences for target and masked conditions by white noise (filled) or time-reversed (open) maskers from Figs. 10 and 11 [7].

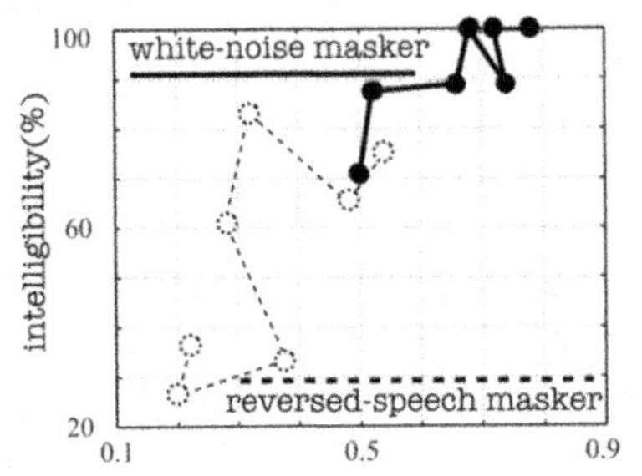

FIGURE 3.12

Intelligibility as function of the 75*th* percentile for the cross-correlation coefficients from Fig. 12 [7].

normalized with unit magnitudes, indicates the distribution of the spacing in the prominent spectral components. The differences in the masking effects between the energy and informational maskers imply that the loss of intelligibility might be related to losing time-dependent harmonic structures as well as the narrowband envelopes under masked conditions.

3.4 Exercises

1. Derive analytic representations for the following waveforms: [1][2]

$$\begin{aligned}&(1)\ \sin x+\sin 2x\ \ (2)\ \cos x+\cos 2x\ \ (3)\ \sin x+2\sin 2x\\&(4)\ \cos x+2\cos 2x\ \ (5)\ \sin x+\cos x\ \ (6)\ \sin x+\cos 2x\\&(7)\ \sin 2x+\cos x\ \ (8)\ \sin x+2\cos 2x\ \ (9)\ 2\sin 2x+\cos x.\end{aligned}\tag{3.33}$$

2. Show the amplitude modulation form for the following superpositions of sinusoidal functions, where $0 < m \le 1$:

$$\begin{aligned}&(1)\ \frac{m}{2}\sin(\omega_c-\Delta\omega)t+\sin\omega_c t+\frac{m}{2}\sin(\omega_c+\Delta\omega)t\\&(2)\ \frac{m}{2}\cos(\omega_c-\Delta\omega)t+\cos\omega_c t+\frac{m}{2}\cos(\omega_c+\Delta\omega)t\\&(3)\ \frac{m}{2}\sin((\omega_c-\Delta\omega)t-\Delta\phi)+\sin\omega_c t+\frac{m}{2}\sin((\omega_c+\Delta\omega)t+\Delta\phi)\\&(4)\ \frac{m}{2}\sin((\omega_c-\Delta\omega)t+\Delta\phi)+\sin\omega_c t+\frac{m}{2}\sin((\omega_c+\Delta\omega)t-\Delta\phi)\\&(5)\ \frac{m}{2}\sin(\omega_c-\Delta\omega)t+\cos\omega_c t+\frac{m}{2}\sin(\omega_c+\Delta\omega)t\\&(6)\ \frac{m}{2}\cos(\omega_c-\Delta\omega)t+\sin\omega_c t+\frac{m}{2}\cos(\omega_c+\Delta\omega)t\end{aligned}\tag{3.34}$$

3. Show the cross-correlation coefficients for the following pairs of random signals:

$$(1)\ X+Y,\ X-Y\ \ (2)\ X,\ aX\ \ (3)\ X+aY,\ X-aY\ \ (4)\ X,\ RX+\sqrt{1-R^2}Y$$

where $\mathrm{E}[X^2]=\mathrm{E}[Y^2]=1$, $\mathrm{E}[X]=\mathrm{E}[Y]=0$, and $0<|R|\le 1$.

4. Rewrite the following equation using the cross-correlation coefficient R_{XY}

$$K=\frac{\mathrm{E}[(X-Y)^2]}{\mathrm{E}[(X+Y)^2]},$$

where $\mathrm{E}[X^2]=\mathrm{E}[Y^2]=1$ and $\mathrm{E}[X]=\mathrm{E}[Y]=0$.

5. Suppose a function $X(t)=X_e(t)+X_o(t)$, where $X_e(t)$ (or $X_o(t)$) denotes an even (or odd) function. Show the cross-correlation between $X(t)$ and $X(-t)$, where $\mathrm{E}[X_e(t)^2]=\mathrm{E}[X_o(t)^2]$ and $\mathrm{E}[X_e(t)]=\mathrm{E}[X_o(t)]=0$.

6. Explain the terminologies listed below:
(1) analytic signal (2) Hilbert envelope (3) Hilbert carrier
(4) instantaneous magnitude (5) instantaneous phase (6) instantaneous frequency
(7) causal signal (8) causal in frequency domain (9) amplitude modulation
(10) modulation index (11) diotic listening condition (12) auto-correlation sequence
(13) spectral auto-correlation sequence (14) informational masker (15) energy masker.

References

[1] M. Tohyama, Sound in the Time Domain, Springer, 2017.
[2] M. Tohyama, Sound and Signals, Springer, 2011.
[3] M. Tohyama, T. Koike, Fundamentals of Acoustic Signal Processing, Academic Press, 1998.
[4] M. Tohyama, Waveform Analysis of Sound, Springer, 2015.
[5] R. Kumaresari, A. Rao, On minimum/maximum/all-pass decomposition in time and frequency domains, IEEE Trans. SP 48 (2000) 2973–2976.
[6] T. Houtgast, H.J.M. Steeneken, R. Plomp, A review of the MTF concept in room acoustics and its use for estimating speech intelligibility in auditoria, J. Acoust. Soc. Am. 77 (3) (1985) 1069–1077.
[7] Y. Hara, M. Tohyama, K. Miyoshi, Effects of temporal and spectral factors of maskers on speech intelligibility, Appl. Acoust. 73 (9) (2012) 893–899.
[8] M.R. Schroeder, S. Mehrgardt, Auditory masking phenomena in the perception of speech, in: R. Carlson, B. Granstroem (Eds.), The Representation of Speech in the Peripheral Auditory System, Elsevier Biomedical, Amsterdam, 1982, pp. 79–87.
[9] R. Lutfy, How much masking in informational masking?, J. Acoust. Soc. Am. 88 (6) (1990) 2607–2610.
[10] Y. Shimizu, M. Fujiwara, Speech privacy in Japan, relevant feeling, evaluation, in: Proceedings of Inter-Noise 2009, Ottawa, 2009.
[11] B.C.J. Moore, An Introduction to the Psychology of Hearing, Academic Press, 1997.
[12] K. Yoshida, M. Kazama, M. Tohyama, Pitch- and speech rate conversion using envelope modulation modeling, in: Int. Conf. Acoustics, Speech, and Signal Processing I, SP-P04.04, 2003, pp. 425–428.
[13] K. Terada, M. Tohyama, T. Houtgast, The effect of envelope or carrier delays on the precedence effect, Acustica 91 (6) (2005) 1016–1019.
[14] Y. Hara, M. Matsumoto, K. Miyoshi, Method for estimating pitch independently from power spectrum envelope for speech and musical signal, Symp. Temporal Des. Arch. Environ. 9 (1) (2009) 121–124.

CHAPTER

Spectral and temporal effects of signals on speech intelligibility

4

CONTENTS

4.1 Frame-wise Fourier transform

4.1.1 N-point Fourier transform

Suppose a discrete sequence $a(n)$. The generating function $A(X)$ is defined for the sequence as [1]

$$A(X) = \sum_{n} a(n)X^n, \tag{4.1}$$

Acoustic Signals and Hearing. https://doi.org/10.1016/B978-0-12-816391-7.00012-7

where X denotes a formal variable. Taking another sequence $b(n)$, the generating function can be defined as

$$B(X) = \sum_n b(n)X^n. \tag{4.2}$$

A product of the generating functions, $C(X)$, generates the new sequence $c(n)$ as

$$A(X)B(X) = C(X) = \sum_n c(n)X^n, \tag{4.3}$$

where $c(n)$ is expressed as

$$c(n) = a * b(n), \tag{4.4}$$

which is called the convolution of a pair of sequences $a(n)$ and $b(n)$ [1]. Substituting Eqs. (4.1) and (4.2) for Eq. (4.3), the sequence $c(n)$ can be rewritten as

$$c(n) = a * b(n) = \sum_m a(m)b(n-m). \tag{4.5}$$

Assuming the formal variable X to be z^{-1}, then

$$A(z^{-1}) = \sum_n a(n)z^{-n}, \tag{4.6}$$

which is called the z-transform of the sequence $a(n)$, where z is a complex variable such that

$$z = |z|\mathrm{e}^{\mathrm{i}\Omega}. \tag{4.7}$$

Here taking $|z| = 1$ as

$$z = \mathrm{e}^{\mathrm{i}\Omega}, \tag{4.8}$$

then

$$A(z^{-1})\Big|_{z=\mathrm{e}^{\mathrm{i}\Omega}} = A(\mathrm{e}^{-\mathrm{i}\Omega}) = \sum_n a(n)\mathrm{e}^{-\mathrm{i}\Omega n} \tag{4.9}$$

is derived. This is called the Fourier transform of the sequence $a(n)$ that is defined on $z = \mathrm{e}^{\mathrm{i}\Omega}$ or on the unit circle in the complex frequency domain.

The Fourier transform of a sequence $a(n)$ given by Eq. (4.9) is a continuous and periodic function of Ω with the period 2π. The sequence $a(n)$ denotes coefficients of the Fourier series expansion of $A(\mathrm{e}^{-\mathrm{i}\Omega})$ as [2][3]

$$\hat{a}(n) = \frac{1}{2\pi}\int_0^{2\pi} A(\mathrm{e}^{-\mathrm{i}\Omega})\mathrm{e}^{\mathrm{i}\Omega n}\,d\Omega. \tag{4.10}$$

Eq. (4.10) is called the representation of a discrete signal $a(n)$ using the continuous and periodic spectral function. The sequence $\hat{a}(n)$ that is obtained by substituting

Eq. (4.9) for Eq. (4.10) should be considered. The difference between $a(n)$ and $\hat{a}(n)$ is given in the Exercises at the end of this chapter.

Sampling the spectral function $A(\mathrm{e}^{-\mathrm{i}\Omega})$ at $\Omega = 2\pi k/N$ $(0 \le k \le N-1)$, the pair of discrete Fourier and inverse Fourier transforms for a discrete sequence $a(n)$ is called the N−point discrete Fourier transform (DFT) pair, which is the short form of [2]

$$A(k) = A(\mathrm{e}^{-\mathrm{i}\Omega})\Big|_{\Omega=\frac{2\pi}{N}k} = \sum_{n=0}^{N-1} a(n)\mathrm{e}^{-\mathrm{i}\frac{2\pi}{N}kn} \tag{4.11}$$

and

$$\overline{a}(n) = \frac{1}{N}\sum_{k=0}^{N-1} A(k)\mathrm{e}^{\mathrm{i}\frac{2\pi}{N}kn}, \tag{4.12}$$

where $\overline{a}(n)$ denotes the periodic extension of $a(n)$ such that

$$\overline{a}(n \pm pN) = a(n) \quad 0 \le n \le N-1 \tag{4.13}$$

and p is a positive integer. The DFT pair expresses a discrete sequence $a(n)$ by using the discrete and periodic spectral sequence (or periodic line spectral sequence), and thus the sequence is extended periodically outside the region where the sequence is originally given. A sequence cannot be predicted outside the original region according to the Fourier transform, except for periodic sequences.

4.1.2 Linear operation of Fourier transform

The Fourier transform is a linear operation. Suppose a sequence, for example, $a(n)$ $(0 \le n \le 4)$ is divided into

$$a(n) = b_1(n) + b_2(n) \quad 0 \le n \le 4, \tag{4.14}$$

where

$$b_1(n) = \begin{cases} a(n) & 0 \le n \le 1 \\ 0 & 2 \le n \le 4 \end{cases} \quad \text{and} \quad b_2(n) = \begin{cases} 0 & 0 \le n \le 1 \\ a(n) & 2 \le n \le 4. \end{cases} \tag{4.15}$$

The Fourier transforms of the sequences are written as

$$A(\mathrm{e}^{-\mathrm{i}\Omega}) = \sum_{n=0}^{4} a(n)\mathrm{e}^{-\mathrm{i}\Omega n}, \; B_1(\mathrm{e}^{-\mathrm{i}\Omega}) = \sum_{n=0}^{1} a(n)\mathrm{e}^{-\mathrm{i}\Omega n}, \; B_2(\mathrm{e}^{-\mathrm{i}\Omega}) = \sum_{n=2}^{4} a(n)\mathrm{e}^{-\mathrm{i}\Omega n}, \tag{4.16}$$

and confirm

$$A(\mathrm{e}^{-\mathrm{i}\Omega}) = B_1(\mathrm{e}^{-\mathrm{i}\Omega}) + B_2(\mathrm{e}^{-\mathrm{i}\Omega}). \tag{4.17}$$

The Fourier transform of a sequence is a linear operation in which the sequence can be divided into subsequences that create the original sequence by concatenation. Fourier transforms for the subsequences are called the frame-wise Fourier transforms.

4.1.3 Discrete Fourier transform and subsequences

The discrete Fourier transform is also a linear operation. Suppose a sequence $a(n)$ is divided into $b_1(n)$ and $b_2(n)$ once more, but the lengths of the sequences are

$$N = N_a = N_{b_1} = N_{b_2} = 5. \tag{4.18}$$

Taking the DFT of $a(n)$, then

$$A(k) = \frac{1}{5}\sum_{n=0}^{4} a(n)e^{-i\frac{2\pi}{5}kn}. \tag{4.19}$$

Similarly, DFTs for the subsequences are given by

$$B_1(k) = \frac{1}{5}\sum_{n=0}^{1} a(n)e^{-i\frac{2\pi}{5}kn} \quad \text{and} \quad B_2(k) = \frac{1}{5}\sum_{n=2}^{4} a(n)e^{-i\frac{2\pi}{5}kn} \tag{4.20}$$

and confirm

$$A(k) = B_1(k) + B_2(k) \quad 0 \le k \le 4. \tag{4.21}$$

An important point in the example of the discrete Fourier transform $A(k)$ is that the length of the sequences $B_1(k)$ and $B_2(k)$ are identical to $N = 5$.

A limit case would be when all the subsequences are composed of a single entry except zeros such that

$$\begin{pmatrix} \mathbf{b}_0^{\mathrm{T}} \\ \vdots \\ \mathbf{b}_4^{\mathrm{T}} \end{pmatrix} = \begin{pmatrix} a(0) & 0 & 0 & 0 & 0 \\ 0 & a(1) & 0 & 0 & 0 \\ 0 & 0 & a(2) & 0 & 0 \\ 0 & 0 & 0 & a(3) & 0 \\ 0 & 0 & 0 & 0 & a(4) \end{pmatrix}. \tag{4.22}$$

The DFT of the sequence $a(n)$ becomes

$$A(k) = B_0(k) + \cdots + B_4(k), \tag{4.23}$$

where

$$B_m(k) = \frac{1}{5}\sum_{n=0}^{4} a(n)\delta(n-m)e^{-i\frac{2\pi}{5}kn} = \frac{1}{5}a(m)e^{-i\frac{2\pi}{5}km} \tag{4.24}$$

and the magnitude and phase of $B_m(k)$ are

$$|B_m(k)| = \frac{1}{5}|a(m)| \quad \text{and} \quad \angle B_m(k) = \begin{cases} -\frac{2\pi}{5}m \cdot k & a(m) \geq 0 \\ -\frac{2\pi}{5}m \cdot k \pm \pi & a(m) < 0 \end{cases} \tag{4.25}$$

for $0 \leq m \leq 4$ and $0 \leq k \leq 4$. The subsequences $\mathbf{b}_0^{\mathrm{T}} \cdots \mathbf{b}_4^{\mathrm{T}}$ have the DFTs in which the magnitudes are independent of k (the frequency parameter), and the phase (angle) is linearly proportional to k. The phase in proportion to the frequency is called the linear phase. The proportion coefficient m corresponds to the time delay of each sample of $a(m)$.

The Fourier transform given by Eq. (4.20) for the subsequence $b_1(n)$ or $b_2(n)$ denotes the interpolated spectral sequence in which the frequency bin representing the spectral sequence is $1/N$, which is equal to that for the original time sequence. Dividing a signal record $x(n)$ with the length N into l frame-wise sequences $x_i(n)$ can be formulated by windowing sequences with the length L as

$$x_i(n) = w_i(n) \cdot x(n) \quad \text{or} \quad X_i(k) = W_i * X(k), \tag{4.26}$$

where $W_i(k)$ and $X_i(k)$ denote the Fourier transforms for $w_i(n)$ and $x_i(n)$, respectively. Summing over the frame-wise spectral sequences,

$$\sum_{i=0}^{l-1} X_i(k) = \sum_{i=0}^{l-1} W_i * X(k) = W * X(k), \tag{4.27}$$

subject to which

$$\sum_{i=0}^{l-1} W_i(k) = W(k) \tag{4.28}$$

gives the Fourier transform of a sequence composed of unit pulses of length $N = lL$.

On the other hand, the frame-wise DFT customarily divides the sequence into subsequences with the record length of L, where $N = lL$ without adding zeros for spectral interpolation and N denotes the length of the original sequence. Again, take an example of $a(n) \quad 0 \leq n \leq 7$ with subsequences $b_1(n)$ and $b_2(n)$, where

$$\mathbf{b}_1 = \begin{pmatrix} a(0) & a(1) & a(2) & a(3) \end{pmatrix}^{\mathrm{T}}, \tag{4.29}$$

$$\mathbf{b}_2 = \begin{pmatrix} a(4) & a(5) & a(6) & a(7) \end{pmatrix}^{\mathrm{T}}. \tag{4.30}$$

The record lengths of $\mathbf{b}_1$ and $\mathbf{b}_2$ are identical to each other as

$$N_{\mathbf{b}_1} = N_{\mathbf{b}_2} = 4, \tag{4.31}$$

while the original sequence has eight entries. Taking the DFTs for the subsequences, then

$$B_1(k) = \frac{1}{4}\sum_{m=0}^{3} a(m)e^{-i\frac{2\pi}{4}km} \text{ and } B_2(k) = \frac{1}{4}\sum_{m=0}^{3} a(m+4)e^{-i\frac{2\pi}{4}k(m+4)}, \tag{4.32}$$

where $B_1(k)$ and $B_2(k)$ yield $\mathbf{b}_1$ and $\mathbf{b}_2$, respectively. The concatenation of $\mathbf{b}_1$ and $\mathbf{b}_2$ represents the original sequence $a(n)$.

The spectral sequences $B_1(k)$ and $B_2(k)$, however, are composed of four spectral entries corresponding to $k = 0, 1, 2, 3$, although the original sequence has eight spectral components $k = 0, 1, 2, \cdots 7$. Spectral fine structures for the subsequences are quite likely lost compared to the original sequence. In contrast temporal information or temporal dependence of spectral components could be clearly presented by the subsequences, while the temporal information would be lost in spectral fine structures derived from the entire original sequence. Defining a frame length or record lengths of subsequences would be an important issue for signal signature analysis such as representation of intelligible speech materials. An informative frame length to represent speech samples would be described from the perspective of intelligibility in this chapter [4].

The difference in the spectral sequences between the original and subsequences is interesting. A sum of the spectral components for the subsequences

$$S_B(k) = \frac{1}{2}(B_1(k) + \hat{B}_2(k)) = \frac{1}{8}\sum_{m-0}^{3}(a(m) + a(m+4))e^{-i\frac{2\pi}{4}km} \tag{4.33}$$

$$\hat{B}_2(k) = \frac{1}{4}\sum_{m=0}^{3} a(m+4)e^{-i\frac{2\pi}{4}km} \tag{4.34}$$

after discarding the linear-phase shift for $b_2(k)$. Compare the spectral sequences between the original and the sum of the subsequences. Interestingly, the sum of the spectral sequences can be written as

$$S_B(0) = \frac{1}{8}\sum_{m=0}^{3}(a(m) + a(m+4)) = \frac{1}{8}\sum_{m=0}^{7} a(m) = A(0) \tag{4.35}$$

$$S_B(1) = \frac{1}{8}\sum_{m=0}^{3}(a(m) + a(m+4))e^{-i\frac{2\pi}{8}2m} \tag{4.36}$$

$$= \frac{1}{8}\sum_{m=0}^{3} a(m)e^{-i\frac{2\pi}{8}2m} + \frac{1}{8}\sum_{m=4}^{7} a(m)e^{-i\frac{2\pi}{8}2m} = A(2)$$

$$S_B(2) = \frac{1}{8}\sum_{m=0}^{3}(a(m) + a(m+4))e^{-i\frac{2\pi}{8}4m} = A(4) \tag{4.37}$$

$$S_B(3) = \frac{1}{8}\sum_{m=0}^{3}(a(m) + a(m+4))e^{-i\frac{2\pi}{8}6m} = A(6). \tag{4.38}$$

The spectral entries for the subsequences correspond (or are identical) to the even number entries for the original sequence. The period of the subsequences reconstructed by their spectral components is 4, while the original is represented by the period 8, where

$$A(k) = \frac{1}{8}\sum_{n=0}^{7} a(n)e^{-i\frac{2\pi}{8}kn}. \tag{4.39}$$

The result of the spectral sequence by $S_B(k)$ is an example of the relationship (or trade off) between the spectral resolution and time resolution. The superposition of the interpolated spectral sequence provides the original spectral sequence for the total length of the time record, as shown in Eq. (4.21). Fig. 4.1 shows an example of the differences in subsequences with and without spectral interpolation.

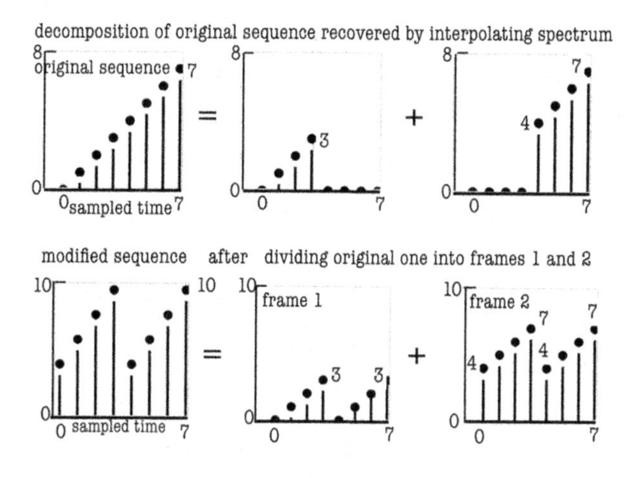

FIGURE 4.1

Examples of subsequences assuming spectral interpolation (upper panel) and no interpolation (lower panel).

The upper panel displays the subsequences assuming the spectral interpolation (or zero-adding). The original time sequence can be restored from the superposition of the two frame-wise spectral subsequences. In contrast, the original time sequence is no longer reconstructed, even after summing over the spectral subsequences in the lower panel because there is no spectral interpolation.

4.1.4 Magnitude and phase effects on waveform

Fourier transforms of real sequences are complex functions that are composed of magnitude and phase spectral components. Spectral sequences composed of the magnitude and the phase components are necessary to represent a real sequence. Magnitude or power spectral components would be understandable recalling that the pitch of a periodic signal is based on the harmonic structure of the signal. Phase spectral behavior might be a more puzzling question than the magnitude spectral property. Looking at some examples would be helpful in order to understand the significance of the phase.

Suppose an impulse whose magnitude or power spectral function is ideally flat (with a linear phase spectral function) independent of the frequency. Randomizing the phase (discarding the linear phase), the reconstructed signal might no longer be the impulse; it looks random noise instead. A similar example would be an artificial reverberator that may be familiar to audio engineers [5]. The reverberator is normally called an all-pass filter in which the magnitude spectral characteristics are flat independent of the frequency, similar to an impulse signal, while the phase could be random. The complexity of the phase spectral characteristics is due to the randomness of the reverberant sound field [6][7].

Another example is amplitude modulation in which three sinusoidal components with slightly different frequencies are contained. Suppose a sum of three sinusoidal components such that (see Exercises 3.4-2)

$$\begin{aligned} y(t) &= \frac{m}{2}\sin(\omega_c - \Delta\omega)t + \sin\omega_c t + \frac{m}{2}\sin(\omega_c + \Delta\omega)t \\ &= (1 + m\cos\Delta\omega t)\sin\omega_c t = e(t)\sin\omega_c t, \end{aligned} \tag{4.40}$$

where

$$e(t) = 1 + m\cos\Delta\omega t \geq 0 \tag{4.41}$$

denotes the envelope, assuming $\omega_c > \Delta\omega$, with the modulation index m ($0 < m \leq 1$). In the amplitude modulation given by Eq. (4.40), $m/2$ represents the magnitude spectra for the sinusoidal components of $\omega_c \pm \Delta\omega$.

The effects of the phase spectra on the waveform are interesting. Take another example:

$$\begin{aligned} y(t) &= \frac{1}{2}\sin\left((\omega_c - \Delta\omega)t + \Delta\theta\right) + \sin\omega_c t + \frac{1}{2}\sin\left((\omega_c + \Delta\omega)t - \Delta\theta\right) \\ &= (1 + \cos(\Delta\omega t + \Delta\theta))\sin\omega_c t. \end{aligned} \tag{4.42}$$

The example in Eq. (4.42) indicates that the effect of the phase spectrum appears on the envelope rather than the carrier. The local behavior of the phase with respect to the frequency can be expressed by the differential function of the phase regarding the frequency. The effect of the phase spectrum on the modulation envelope in Eq. (4.42) can be rewritten as

$$y(t) = \left(1 + \cos\Delta\omega(t - \tau_g)\right)\sin\omega_c t, \tag{4.43}$$

where

$$\tau_g = \frac{-d\theta}{d\omega} \tag{4.44}$$

is called the group delay (s). When the group delay is positive (negative), the envelope is delayed (advanced).

An irregular case when the phase is not a smooth function or not differentiable would be intriguing from the perspective of the modulation envelope. Take an example such as

$$\begin{aligned} y(t) &= \frac{1}{2}\sin\left((\omega_c - \Delta\omega)t + \frac{\pi}{2}\right) + \sin\omega_c t + \frac{1}{2}\sin\left((\omega_c + \Delta\omega)t + \frac{\pi}{2}\right) \\ &= \sqrt{1+\cos^2\Delta\omega t}\cos(\omega_c t - \theta) = e(t)\cos(\omega_c t - \theta) \end{aligned} \tag{4.45}$$

that indicates $\pi/2$ phase shift at the carrier frequency to both of the side frequencies, where

$$\cos\theta = \frac{\cos\Delta\omega t}{\sqrt{1+\cos^2\Delta\omega t}} \quad \text{and} \quad e(t) = \sqrt{1+\cos^2\Delta\omega t}. \tag{4.46}$$

The envelope $e(t)$ does not exemplify the amplitude modulation well because

$$1 \le \sqrt{1+\cos^2\Delta\omega t} \le 2, \tag{4.47}$$

and thus no perfect modulation is realized [8]. The waveform of Eq. (4.45) is illustrated in Fig. 4.2.

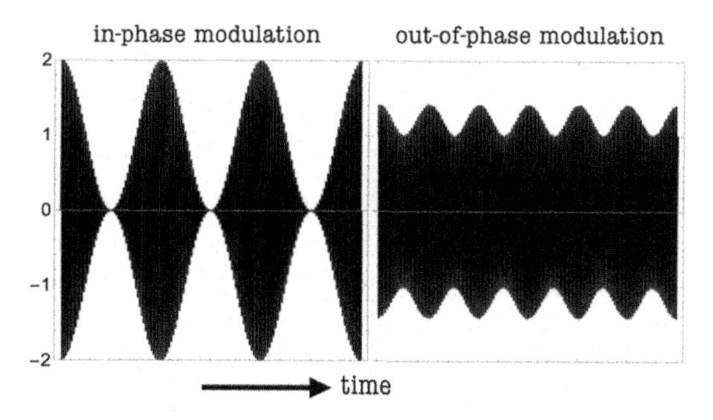

FIGURE 4.2

Perfect (left) and nonperfect (right) modulations under in-phase (left) and out-of-phase carrier conditions.

Amplitude modulation waveforms are sensitive not only to the modulation index (or the magnitude spectral condition), but also to the phase relationship between the carrier and its side components. The left panel in Fig. 4.2 shows perfect modulation under the in-phase condition between the side-frequency and the carrier components. In contrast, the right panel presents the waveform under the out-of-phase condition between the carrier and side components (no linear phase relation in the three sinusoidal components). No perfect modulation can be seen in the right panel due to the phase effect even under the modulation index is unity. The phase spectral effects on the amplitude modulation would be remarkable from a perceptual standpoint.

4.1.5 Envelope and frame-wise Fourier transforms

Envelopes are formally expressed in a product of two functions such as beats made by a pair of sinusoids. Take an example of a random noise multiplied by a sinusoidal function such that

$$y(t) = s(t)n(t) = \cos\omega_0 t \cdot n(t), \tag{4.48}$$

where $n(t)$ is a random noise, and the absolute of $s(t)$ is assumed to be the envelope. Following Eq. (4.26), the Fourier transform of $y(t)$ can be expressed as

$$Y(\omega) = S * N(\omega), \tag{4.49}$$

where $S(\omega)$ and $N(\omega)$ denote the Fourier transforms of $s(t)$ and $n(t)$, respectively. The convolution in the frequency domain recalls the phase-spectral correlation such that [4]

$$R_{ph}(\Delta k) = \mathrm{E}\left[\mathrm{e}^{\mathrm{i}(\phi(k+\Delta k)-\phi(k))}\right], \tag{4.50}$$

where E denotes the ensemble average according to the definition of the correlation for random events, $\phi(k)$ gives the phase spectrum at the frequency k, and k gives the number of line spectral components. Recalling a product of two sinusoidal components

$$\cos\Delta\omega t \cdot \cos\omega_c t = \frac{1}{2}(\cos(\omega_c + \Delta\omega t) + \cos(\omega_c - \Delta\omega t)) \tag{4.51}$$

that yields $\omega_c \pm \Delta\omega$ components, the phase correlation is expected at $2\Delta\omega$ spacing for modulated noise.

Fig. 4.3 shows a sample of the waveform given by Eq. (4.48) with its magnitude and phase spectral function of the Fourier transform. Looking at the waveform in the right-top panel, a modulated envelope can be clearly seen. On the other hand, the magnitude and/or the phase spectral figures may not be very informative when estimating the envelope.

An example of the absolute R_{ph} is displayed in Fig. 4.4 comparing the random noise with and without modulation. As expected, the auto-correlation of the phase for the random noise looks like an impulse at $\Delta k = 0$ similar to white noise. In contrast, the spacing for the phase correlation of the modulated noise gives the frequency of the envelope.

The correlations in the phase show that the phase spectrum might reconstruct the envelope even after the magnitude spectrum is replaced by that for the random noise without the modulation envelope. Fig. 4.5 illustrates an example of the modulated noise reconstructed by using its phase with random magnitude spectrum. The reconstructed envelope can be clearly seen.

The phase correlation produces the difference between modulated and nonmodulated random noise. The difference in the phase correlation is due to differences in the fine structure of the phase spectrum. Frame-wise Fourier transforms with relatively short frame lengths do not provide the fine resolution of the frequency. Thus, the reconstruction or representation of the modulation envelope using the phase spectrum

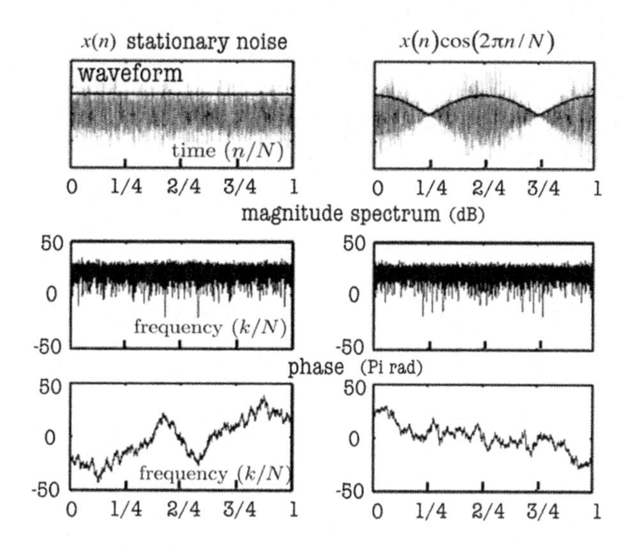

FIGURE 4.3

Waveforms of random noise with and without modulation, and their magnitude and phase spectral characteristics from Fig. 5 [4].

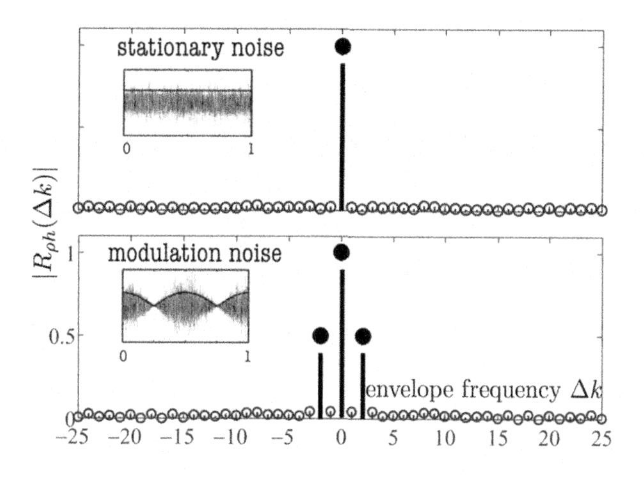

FIGURE 4.4

Examples of magnitude of phase auto-correlation for random noise with and without modulation envelope shown in Fig. 4.3 from Fig. 6 [4].

may be possible under long frame length conditions. Interestingly, the loss of the frequency resolution by frame-wise Fourier transforms could be compensated by time resolution in the time domain for modulated envelopes. Frame-wise Fourier transforms may provide the time resolution under the short frame length condition instead of the frequency resolution.

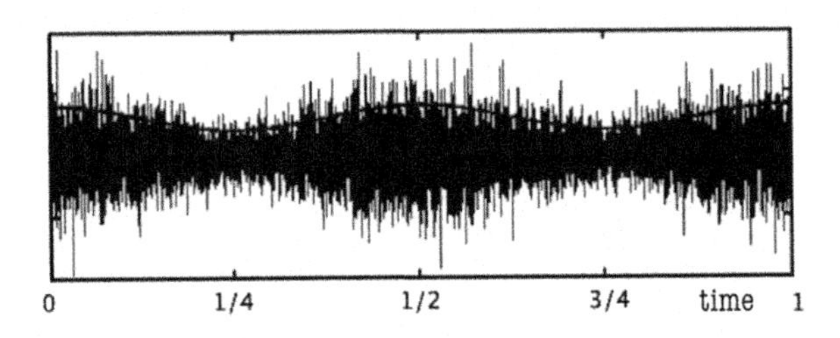

FIGURE 4.5

Reconstructed envelope for modulated noise by phase spectrum of the original modulated noise, but with the random magnitude spectrum from Fig. 7 [4].

Fig. 4.6 represents reconstruction of the envelope of the modulated noise by using the magnitude spectra with random phase.

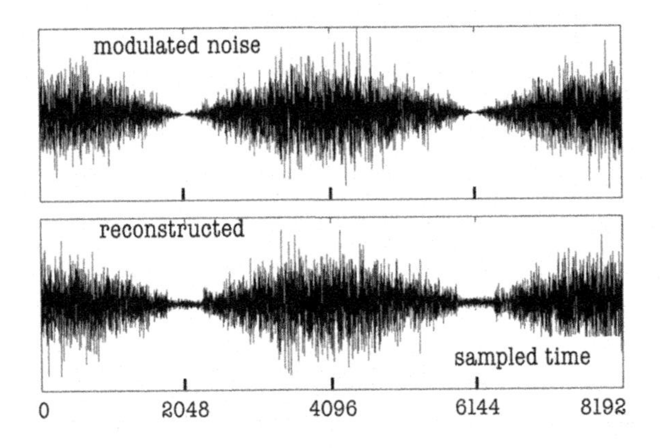

FIGURE 4.6

Representation of envelope by magnitude spectra of frame-wise Fourier transforms (frame length: 256) with random phase. The upper panel is the original waveform and the lower panel is reconstructed from Fig. 3.5 [2].

The frame-lengths for the frame-wise Fourier transforms are 256 samples in the example; the length is sufficiently shorter than the period of the envelope. The modulation envelope can be mostly represented by the frame-wise magnitude spectra with random phase. The envelope, which is conveyed in the fine structure of the phase spectrum, can be represented by temporally (or frame-wise) changing the magnitude spectra, even after the phase spectra are replaced by random phase.

Recalling that phase effects can be observed by amplitude modulation, as shown by Eq. (4.42), temporally changing magnitude spectral analysis by frame-wise Fourier transforms implies a possible way in which the hearing organ perceives the phase effect on the waveform in the time domain [3][9][10]. Speech intelligibility, however, mostly depends on the narrowband envelopes rather than the entire envelope [11]. Representation of intelligible speech envelopes would be a great issue for signature analysis from a perceptual frame of reference [4].

4.2 Intelligibility of synthesized speech and noise spectral components

4.2.1 Frame-wise swapping of magnitude and phase spectra between speech and random noise

Reconstruction of envelopes by using spectral components depends on the frame length under which the magnitude and phase spectra are obtained. Intelligibility of speech materials is closely related to narrowband envelopes of the waveforms of speech samples [11]. An interesting question would be which is significant for representation of intelligible speech the magnitude or phase spectral components. The significance in the intelligibility may be frame length dependent. A frame-wise swapping scheme as shown in Fig. 4.9 may be informative [4][12].

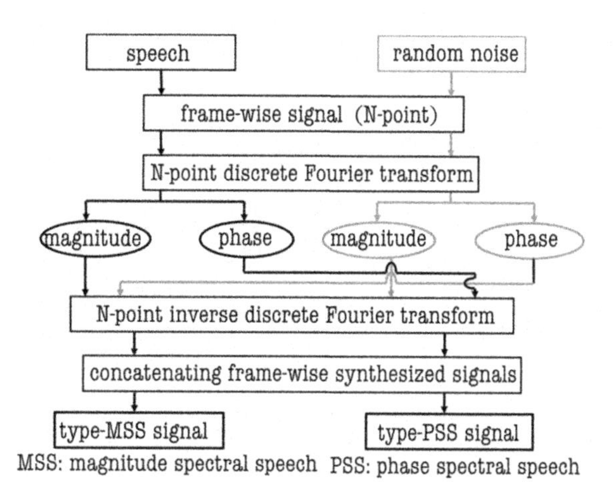

FIGURE 4.7

Frame-wise spectral swapping between intelligible speech and random noise from Fig. 1 [4].

In the experimental scheme, an intelligible speech sentence and random noise, such as white noise sample, are used. The two sound samples are simultaneously cut into frames with under 50% overlapped rectangular windowing except for the two-point or the single-point frames. After taking the Fourier transform in every frame, the magnitude (or phase) spectral components of the speech and noise materials are swapped with each other; these pairs of speech-magnitude with noise-phase, and speech-phase with noise-magnitude are made in each frame. Two hybrid or swapped signals are synthesized through the inverse Fourier transform in each frame. A triangular window with a frame length identical to the rectangular window used for the Fourier transform is applied to every synthesizing frame to avoid discontinuities between successive frames. In Fig. 4.7 MSS (PSS) denotes the synthesized signal using a pair of speech-magnitude (speech-phase) and noise-phase (noise-magnitude). Details of the signal processing conditions can be seen in reference [4].

4.2.2 Sentence intelligibility scores

Synthesized speech materials were subjectively evaluated in terms of sentence intelligibility [4]. The upper panel of Fig. 4.8 shows the sentence intelligibility scores with the standard deviation as functions of the frame length.

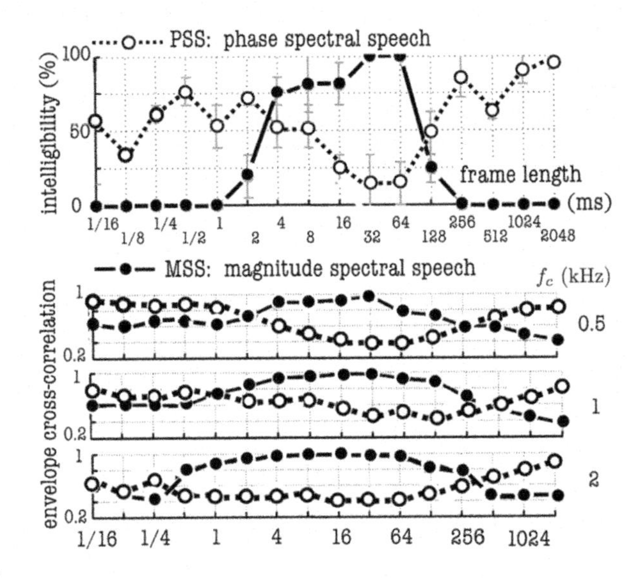

FIGURE 4.8

Intelligibility scores and envelope cross-correlation coefficients. Upper panel: Sentence intelligibility scores for synthesized speech materials MSS and PSS from Figs. 2 [4]. Lower panels: Cross-correlation coefficients of squared narrowband envelopes between original and synthesized (MSS or PSS) envelopes in 1/4–octave bands at center frequencies f_c of 0.5, 1.0, and 2 (kHz) from Fig. 4 [4].

Each intelligibility score in the upper panel is estimated according to six presentations to seven listeners at every frame length condition. A score of 100% indicates that all subjects could correctly write down a sentence without standard deviation.

The scores in the upper panel depend on the frame length condition, as expected, but differently for MSS (magnitude spectral speech with random phase) and PSS (phase spectral speech with random magnitude). The scores for MSS dramatically reveal the frame length dependence ranging from completely unintelligible (for $1/16 - 1$ (ms) and $256 - 2048$ (ms)) to almost perfectly intelligible (for $4 - 64$ (ms) frame length conditions). The almost perfect intelligibility for MSS at around 30 (ms) conditions may explain why only the magnitude spectral components are taken by discarding the phase in conventional speech analysis.

On the other hand, the intelligibility is mostly lost for frame length conditions longer than 256 (ms) [13]. It should be remembered that in room reverberation or artificial reverberation [5] the phase spectrum is mostly random, even if the magnitude spectrum could be flat. Take a longer frame length such as 256 (ms) or longer, so the spectral characteristics of MSS could be similar to that for reverberation sound. Very low intelligibility of reverberation would be intuitively understandable. The MSS actually sounds very reverberant under the longer frame length conditions. This intelligibility loss as the frame length becomes longer than around 128 (ms) shows the loss of time resolution for reconstructing envelopes because of frame lengths that are too long. Time resolution finer than the period of the envelope is necessary, at least, as shown in Fig. 4.8. The frame length 128 (ms) at the breaking point of the intelligibility for MSS indicates the period of envelopes responsible for the intelligibility might be shorter than around 128 (ms).

The results would be intriguing for MSS under frame lengths shorter than 1 (ms). The frequency (or time) resolution for frame-wise Fourier transforms becomes finer (or lost) as the frame length is longer. On the other hand, resolution of the frequency (or time) is lost (or finer) as the frame length becomes shorter. The time resolution is significant to represent the envelope of a waveform by using the frame-wise magnitude spectral records. The frame lengths shorter than 1 (ms) seem to be sufficient to represent the envelope. However, recall that narrowband envelopes, which are defined as envelopes for narrowband (not wideband) waveforms, are significant in representing intelligible speech [11]. It seems quite unlikely under the condition of frame length shorter than 1 (ms) to reconstruct the frequency-band-dependent envelopes. The intelligibility score for MSS implies that a frequency resolution finer than about 250 (Hz) or a frame length longer than 4 (ms) might be necessary to represent the frequency-band-dependent envelopes of intelligible speech by using frame-wise magnitude spectra with random phase.

In contrast to MSS, the results of PSS are mostly less extreme. Recalling that the phase correlation creates the envelope of amplitude-modulated noise, as shown in Fig. 4.5, the fine spectral resolution obtained by frame length conditions longer than 256 (ms) possibly reconstructs envelopes that are highly intelligible. The breaking point of PSS is also found around the frame length of 128 (ms). This frame length condition interestingly implies compensation of the time (or frequency) resolution loss for MSS (or PSS) by fine frequency (or time) resolution for PSS (or MSS).

Another breaking point for shorter frame length conditions could be found at a frame length of around 4 (ms). The compensation of the frequency (time) resolution loss for MSS (or PSS) could be filled in by the time (or frequency) resolution for PSS (or MSS). Preserved intelligibility for PSS would be intriguing at shorter frame lengths than 4 (ms). The limit case when a single point Fourier transform is taken at each frame gives the zero-crossings for PSS because the sign (positive or negative) is kept for PSS, as shown by Eq. (4.25). Zeros-crossing waves made from speech materials could be intelligible, although it might not be perfect. The envelopes could be partly reconstructed by PSS even in the shorter frame length conditions.

4.3 Envelopes of synthesized speech material

4.3.1 Cross-correlation between synthesized and original envelopes

Intelligibility may be closely related to preserving narrowband envelopes [11]. The trends for the envelope recovery with respect to the frame length conditions might follow the intelligibility scores. The cross-correlation coefficient may be an estimate of similarity of envelopes between the original and synthesized envelopes. Details of the envelope correlation analysis is developed in the reference [4].

The lower panels in Fig. 4.8 presents the cross-correlation coefficients between the original and synthesized envelopes; the trends of the correlation mostly follow the intelligibility scores for MSS and PSS, respectively. The breaking point mentioned in Section 4.2.2 seems to shift toward around 256 – 512 (ms) frame length condition (longer than 128 (ms) read from the intelligibility scores) almost independent of the frequency bands. This longer frame length condition estimated by the correlation coefficients implies slower modulation in the envelopes might not be helpful in improving the intelligibility [4].

Interestingly, the other breaking point seems frequency dependent. The frequency resolution takes over the time resolution at the breaking point. After dividing the frequency range into narrowbands, the time resolution required for every frequency band may be frequency dependent. The time resolution needed to compensate for the frequency resolution loss may be shorter, as the center frequency of the narrowband goers higher (or the frequency band becomes wider). The breaking points in the panels actually come to shorter frame lengths, and the points are found around the period of the center frequency for every frequency band.

Magnitude or phase spectral dominance on speech intelligibility could be frame length dependent from the perspective of synthesizing narrowband envelopes by using frame-wise Fourier transforms. Phase spectral properties have received less attention; however, synthesized speech materials show the phase dominance in shorter or longer frame length conditions, while the magnitude spectral dominance could be confirmed in the medium frame length conditions where most speech processing is performed. The phase dominance under the shorter frame length conditions would partly explain why zero-crossing speech might not be perfect, although it is intelligible.

4.3.2 Phase dominance and preserving zero-crossings

The phase dominance on speech intelligibility rather than the magnitude spectrum can be seen under both of the shorter and longer frame lengths conditions. Frequency resolution fine enough to represent the phase correlation creates the envelopes from the phase spectral components. On the other hand, fine time resolution renders the envelopes from the phase as well. The limit case of the latter condition is a single point Fourier transform, as shown by Eq. (4.25), in which zero-crossings of the waveform may be preserved by the phase spectrum [4].

Setting the magnitude spectrum to unity instead of that for random noise for simplicity, the waveform under the single point of Fourier transform can be expressed as a rectangular wave. Take an example of waveforms

$$y(t) = e(t)c(t) = \cos t \cdot \cos 6t \tag{4.52}$$

as a modulated signal with an envelope. Preserving the zero-crossings of the waveform and setting the magnitude to unity, the Fourier transforms are shown in Fig. 4.9.

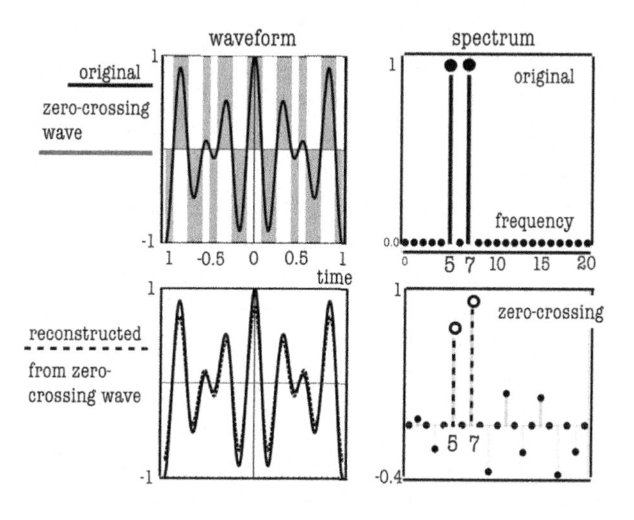

FIGURE 4.9

Sample of modulated waveforms from Eq. (4.52). Top left: waveforms for original and zero-crossing waves, top right: spectral sequence (normalized) of original waveform, bottom right: part of spectral sequence for zero-crossing waveform, and bottom left: reconstructed waveform from zero-crossing by using only its spectral record within a prominent narrowband (frequency numbers 5 and 7) where the original is shown for reference.

Comparing the original and zero-crossing with unit magnitude waves, the spectral components are expanded into the higher harmonics because of the rectified waveform. Performing a narrowband selection corresponding to the sub-band analysis (as shown by frequencies 5 and 7 in Fig. 4.9), the original envelope is possibly reconstructed, unless the modulation frequency is within the width of the frequency band. Coming back to Fig. 4.8, the correlation coefficients decrease for PSS envelopes at shorter frame length conditions (less than $1/4$ (ms)) as the center frequency of the sub-band goes high. This decrease may be partly explained in terms of the width of the frequency band for the recovery of the envelope from the zero-crossings. In principle, the envelopes may be reconstructed by sub-band filtering (which is natural for the hearing organ), of the zero-crossing waveforms, following the spectral relationship between the envelope and carrier, if their prominent spectral components are found in a sub-band.

4.4 Intelligibility of time-reversing or random permutation of speech

4.4.1 Time-reversing a waveform and its Fourier transform

The effect of phase spectral characteristics on speech intelligibility is described in the previous sections. In particular the effect of randomizing the phase on the intelligibility was considered. Time-reversing a waveform in the time domain may be another extremum in the deterioration of the phase spectral behavior from position of intelligibility [14]. Time-reversed speech waveforms are ideal informational maskers rather than energy maskers [15][16][17][18], as described in Chapter 3.

Suppose a waveform $x(t)$ as a real function. The Fourier transform of the function

$$X(\omega) = \frac{1}{2\pi} \int x(t) \mathrm{e}^{-\mathrm{i}\omega t} dt = X_r(\omega) + \mathrm{i} X_i(\omega) \tag{4.53}$$

indicates the original waveform $x(t)$ can be reconstructed through the inverse Fourier transform, where $X_r(\omega)$ and $X_i(\omega)$ denote the real and imaginary parts of $X(\omega)$, respectively. Taking the complex conjugate of $X(\omega)$, then

$$X^*(\omega) = \frac{1}{2\pi} \int x(t) \mathrm{e}^{\mathrm{i}\omega t} dt = X_r(\omega) - \mathrm{i} X_i(\omega) \tag{4.54}$$

is obtained, and

$$\int X^*(\omega) \mathrm{e}^{\mathrm{i}\omega t} d\omega = x(-t). \tag{4.55}$$

Eq. (4.55) indicates that time-reversing a waveform yields a spectrum that is the complex conjugate of the original waveform.

Time-reversing a waveform makes a sentence almost perfectly unintelligible because of phase spectral effects of the complex conjugate spectrum, although the magnitude spectrum is perfectly preserved. With an even or odd waveform, if possible in a speech waveform, the intelligibility may be perfect even after time-reversing the waveform. The Fourier transform of a real and even function is a real and even function that makes no change after time-reversing the waveform. The spectral function of a real and odd waveform is purely imaginary and odd. In addition its signs change the time-reversing; however, the sign change does not affect the intelligibility. The time-reversing effect on symmetric waveforms would be understandable by observing the waveforms because of their symmetry in the time domain.

A speech waveform is neither even nor odd in general. A waveform could be a mixture of the even and odd components. The intelligibility loss by the time-reversing process depends on the mixing condition of the symmetric property of the speech waveform. A question may arise as to what extent a speech waveform is symmetric. The answer would lead to an idea that the symmetry might be frame length dependent under observation conditions.

Assuming the frame length dependent symmetry of a speech waveform, the same type of intelligibility test as that performed in section 4.2 [14] may be possible. A frame-wise time-reversing process is not a linear process. It is not possible to preserve the phase or the magnitude spectrum for the entire waveform after the frame-wise time-reversing. The effect of the frame-wise time-reversing on the intelligibility may depend on the frame length, and generally intelligibility is lost as the frame length increases [19]. The intelligibility loss would be an intriguing issue under shorter frame length conditions.

4.4.2 Intelligibility tests of time-reversing or random permutation of speech

Time-reversing is an extremum in permutation of time samples of a waveform. Another extremum may be random permutation of the speech samples. Comparison in the intelligibility of the two extrema would be interesting from the standpoint of the phase spectral effects on the intelligibility. Fig. 4.10 illustrates synthesizing the test materials where two extrema of the permutation are performed on a frame-wise basis, and the intelligibility scores are organized as a function of the frame length [14][19].

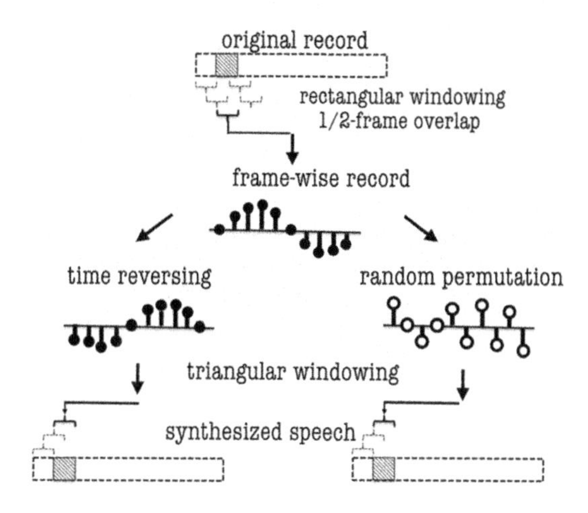

FIGURE 4.10

Frame-wise synthesizing speech materials by time-reversing or random permutation from Fig. 1 [14].

The speech materials are taken from daily Japanese sentences [20] and Harvard sentences of English [21]. A set of 132 processed speech materials spoken by a single female speaker was used in the intelligibility tests of time-reversing or random permutation of speech for Japanese and English, respectively. The length of a single utterance is around 2 (s).

The permutation process is performed frame by frame, where every frame is made by rectangular windowing the original sentence. A half-frame length overlapped triangular window is used in concatenating the processed frames, so that the inter-frame discontinuity might be avoided between a pair of adjacent frames. The 22 frame length conditions are selected from $1/8 - 2048$ (ms) for time-reversing; however, the range of the frame lengths is only up to 8 (ms) for the random permutation as a reference to be compared. To synthesize processed speech materials, 6 original sentences are used in every frame length condition, namely, $132 = 6 \times 22$.

4.4.3 Intelligibility as a function of frame length for time-reversing or random permutation

Six native Japanese (English) listeners took the intelligibility test that is described in detail in the reference [14]. Fig. 4.11 presents the intelligibility scores (averages with standard deviations) for frame-wise time-reversed spoken sentences as a function of the frame length.

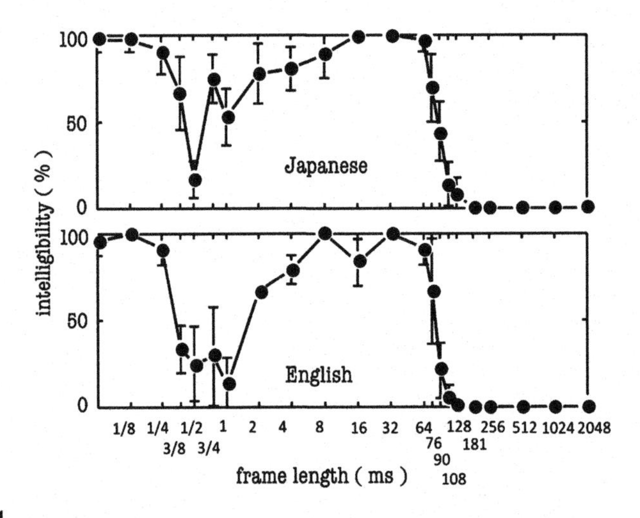

FIGURE 4.11

Sentence intelligibility scores for frame-wise time-reversing sentences; upper: Japanese, lower: English from Fig. 2 [14].

The upper (Japanese) and lower (English) panels are similar to each other.

Very interestingly, the results shown by the two panels are quite similar to the results shown in the top panel in Fig. 4.8. The intelligibility is perfectly lost under frame lengths longer than $128 - 256$ (ms), while synthesized materials preserve the intelligibility even after time-reversing under frame lengths shorter than $1/2 - 1$ (ms). The phase spectral effects on the loss of intelligibility due to time-reversing are prominent for the longer frame lengths. In contrast, the symmetry of the waveforms, which keeps the waveforms even after time-reversing, could be preserved under frame length conditions shorter than $1/2 - 1$ (ms), so that the speech materials might be still intelligible.

The dominance of the magnitude spectrum holds in the middle range of frame lengths. The intelligibility increases as the frame length becomes longer, in which the magnitude spectrum is preserved after time-reversing until the frame length exceeds the range of magnitude spectral dominance. Taking over the dominances of the magnitude or phase spectrum as a function of the frame length may be important again.

On the other hand, Fig. 4.12 displays the results for random permutation [14].

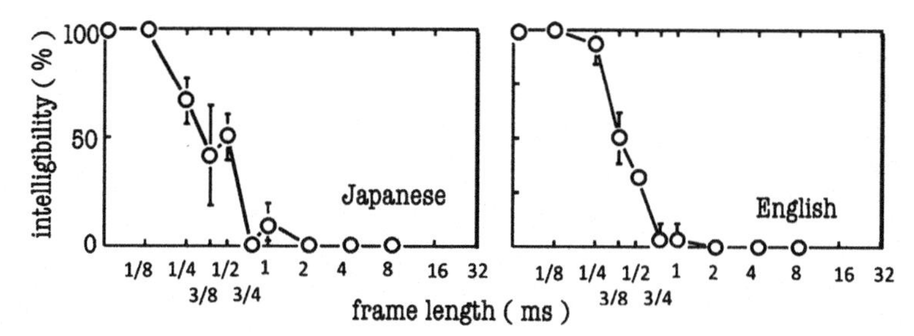

FIGURE 4.12

Similar to Fig. 4.11, but for frame-wise random permutation; left: Japanese, right: English from Fig. 3 [14].

Again no significant differences can be seen between Japanese (left) and English (right); however, the intelligibility loss as a function of the frame length is very different from time-reversing. Neither the magnitude nor the phase spectral property is preserved after the random permutation frame-by-frame. This randomization yields the difference from the time-reversing, and the intelligibility is rapidly lost as the frame length becomes longer. Recovering narrowband envelopes explains the intelligibility, as developed in the next subsection.

4.4.4 Narrowband envelopes of synthesized materials

Recovering narrowband envelopes explains the intelligibility again [4][11][14]. Fig. 4.13 displays examples of the $1/4$–octave band envelopes centered at 1414 (Hz) after time-reversing (left panel) and random permutation (right one) of Japanese sentences under the frame lengths of $1/8$, $1/2$, and 4 (ms) [14].

The envelope of time-reversed material (left panel) is quite similar to the original at $1/8$ (ms) frame length condition, while the similarity breaks down at $1/2$ (ms) frame length; interestingly, the similarity comes back at the condition for 4 (ms) frame length. This similarity and dissimilarity may explain the intelligibility that is dependent on the frame length.

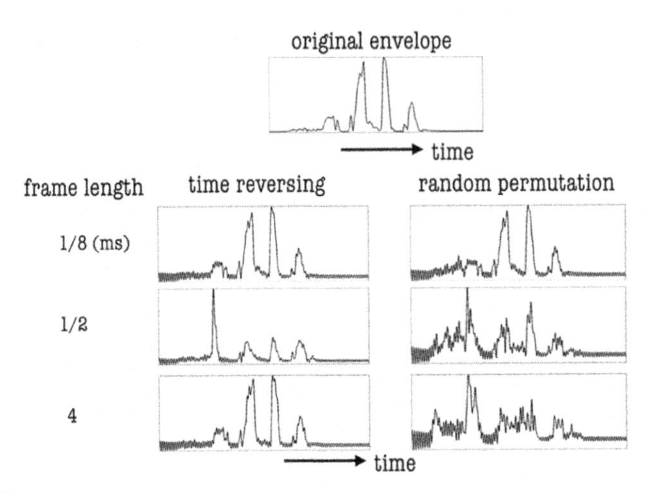

FIGURE 4.13

1/4–octave band envelopes of synthesized materials by frame-wise time-reversing (left panel) and random permutation (right one) from Figs. 4 and 5 [14].

In contrast, the right panel shows envelopes similar to the left panel, but for the materials under the random permutation [14]. The similarity in the synthesized and original envelopes is no longer prominent. The dissimilarity could explain the loss of intelligibility by random permutation. The envelope-similarity or dissimilarity might be expressed by the cross-correlation coefficients between the synthesized and original envelopes [4]. The cross-correlation analysis is described in the next subsection [14].

4.4.5 Cross-correlation of envelopes between synthesized and original materials

A waveform can be represented by a mixture of the even and odd functions. The cross-correlation coefficient between the original and time-reversed envelopes becomes

$$R = \frac{1 \cdot \overline{x_e^2(t)} + (-1) \cdot \overline{x_o^2(t)}}{\overline{x_e^2(t)} + \overline{x_o^2(t)}} \tag{4.56}$$

in general, which indicates a square-based average of $+1$ and -1, where ± 1 means the cross-correlation coefficient for the even or odd component [3]. The magnitudes of cross-correlation coefficients are unity if the envelopes are even or odd; in other words, there is symmetry. On the other hand, the correlation becomes 0 when $\overline{x_e^2(t)} = \overline{x_o^2(t)}$, where $\overline{*}$ gives the long time average. The magnitude or square of the correlation for the time-reversed envelopes gives an estimate of symmetry of the envelopes.

Fig. 4.14 shows squares of the cross-correlation coefficients between the original and frame-wise time-reversed envelopes in the 1/4−octave band centered at 1414 (Hz) [14].

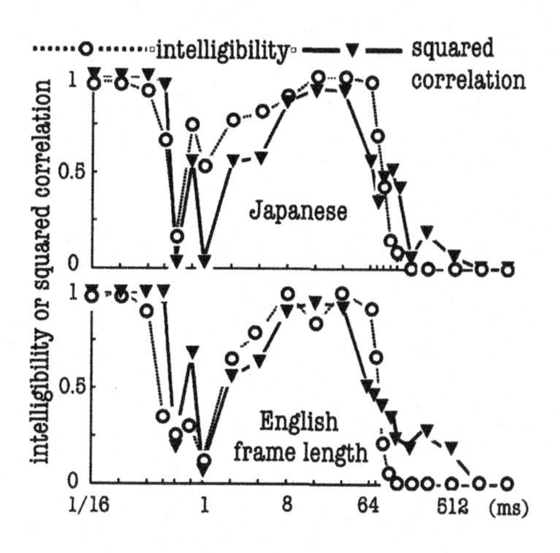

FIGURE 4.14

Squares of cross-correlation coefficients of 1/4−octave band envelopes at 1414 (Hz) for frame-wise time-reversed speech from Fig. 6 [14].

The cross-correlation coefficients are taken as the average for six sentence pairs (used in the experiments) at every frame. The correlations displayed by triangular plots mostly follow the intelligibility scores given by open circles for the upper (Japanese) and lower (English) panels.

The high correlation coefficients for the frame lengths shorter than around 1 (ms) may be due to the symmetry of the waveform or its envelope. In contrast the very low correlation data under the frame lengths longer than around 256 (ms) suggest asymmetry of the waveform.

Interestingly, the recovery of envelope correlations in the medium range corresponds to the magnitude spectral dominance on the narrowband envelope recovery [4]. The symmetry of the waveform is not expected in the medium frame lengths; however, the frame-wise magnitude spectrum keeps the original envelope in every frame even after frame-wise time-reversing. The preservation of frame-wise magnitude spectrum, in other words, recovers the original narrowband envelopes independent of the waveforms unless the frame length meets the time-frequency resolution for speech envelopes recovery [4]. Fig. 4.15 indicates that the intelligibility and the correlation coefficients both rapidly decrease by frame-wise random permutation as the frame length becomes long [14].

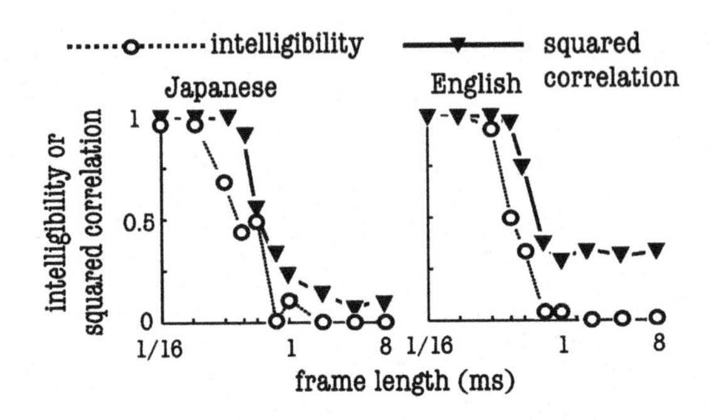

FIGURE 4.15

Similar to Fig. 4.14, but for frame-wise random permutation from Fig. 7 [14].

4.5 Exercises

1. What is the difference between $\hat{a}(n)$ given by Eq. (4.10) and the original sequence $a(n)$ defined for $0 \leq n \leq N-1$.

2. Confirm the discrete Fourier transform pair given by Eqs. (4.11) and (4.12).

3. Confirm $S_B(k)$ in Eq. (4.38).

4. Suppose an observation record with a length of N for a random sound. The spectral data estimated by the Fourier transform of the observation record are random events. What is the relationship between spectral variance and frequency resolution?

5. Derive the Fourier transform for the sequence $y(n)$ [22]

$$y(n) = w(n) \cdot x(n),$$

and the inverse Fourier transform of

$$Y(k) = W * X(k),$$

where $W(k)$ and $X(k)$ are the Fourier transforms of $w(n)$ and $x(n)$, respectively.

6. Derive Eqs. (4.41) and (4.45).

7. A modulating envelope can be estimated by the frame-wise magnitude spectrum or the long-term phase spectral record, as shown in Figs. 4.5 and 4.6. This implies that the envelope or signal dynamics in the time domain can be perceived by hearing according to the frame-wise magnitude spectral behavior of the signal. What is the condition of the frame length for the frame-wise magnitude spectral analysis when the envelopes are frequency dependent? An example of narrowband envelope reconstruction can be seen in Fig. 4.8 for intelligible speech.

8. Show the difference in the spectrum between a sinusoidal wave and its zero-crossing.

9. Confirm Eq. (4.56).

10. Show the magnitude or phase spectral dominance on speech intelligibility as a function of the frame length in Figs. 4.8 and 4.11.

11. Explain terms listed below:

(1) generating function (2) convolution (3) $z-$transform
(4) Fourier transform of a sequence (5) discrete Fourier transform of a sequence
(6) frame-wise Fourier transform
(7) interpolated spectrum (8) group delay (9) envelope
(10) carrier (11) amplitude modulation (12) modulation index
(13) in-phase modulation (14) out-of-phase modulation (15) cross-correlation coefficient
(16) zero-crossing wave (17) magnitude spectrum of time-reversed wave (18) phase-spectrum of time reversed wave

References

[1] R. Nelson, Probability, Stochastic Processes, and Queueing Theory, Springer, 1995.
[2] M. Tohyama, Waveform Analysis of Sound, Springer, 2015.
[3] M. Tohyama, Sound in the Time Domain, Springer, 2017.
[4] M. Kazama, S. Gotoh, M. Tohyama, T. Houtgast, On the significance of phase in the short term Fourier spectrum for speech intelligibility, J. Acoust. Soc. Am. 127 (3) (2010) 1432–1439.
[5] M.R. Schroeder, Improved quasi-stereophony and colorless artificial reverberation, J. Acoust. Soc. Am. 33 (1961) 1061–1064.
[6] M.R. Schroeder, Statistical parameters of the frequency response curves in large rooms, J. Audio Eng. Soc. 35 (5) (1987) 299–306.
[7] M. Tohyama, R.H. Lyon, T. Koike, Reverberant phase in a room and zeros in the complex frequency plane, J. Acoust. Soc. Am. 89 (4) (1991) 1701–1707.
[8] W.M. Hartmann, Signals, Sound, and Sensation, Springer, 1997.
[9] S. Ushiyama, M. Tohyama, M. Izuka, Y. Hirata, Generalized harmonic analysis of non-stationary waveforms (in Japanese with English abstract), Technical Report of IEICE EA-93-103, Inst. Elect. Info., and Comm. Engineers, Japan, 1994-03.
[10] T. Terada, H. Nakajima, M. Tohyama, Y. Hirata, Non-stationary waveform analysis and synthesis using generalized harmonic analysis, in: Proceedings of IEEE-SP. Int. Symp. on Time-Frequency and Time-Scale Analysis, 1994, pp. 429–432.
[11] T. Houtgast, H.J.M. Steeneken, R. Plomp, A review of the MTF concept in room acoustics and its use for estimating speech intelligibility in auditoria, J. Acoust. Soc. Am. 77 (3) (1985) 1069–1077.
[12] L. Liu, J. He, G. Palm, Effects of phase on the perception of intervocalic stop consonants, Speech Commun. 22 (1997) 403–417.
[13] A. Oppenheim, J. Lim, The importance of phase in signals, Proc. IEEE 69 (1981) 529–541.
[14] S. Gotoh, M. Tohyama, T. Houtgast, The effect of permutations of time samples in the speech waveform on intelligibility, J. Acoust. Soc. Am. 142 (1) (2017) 249–255.

[15] M.R. Schroeder, S. Mehrgardt, Auditory masking phenomena in the perception of speech, in: R. Carlson, B. Granstroem (Eds.), The Representation of Speech in the Peripheral Auditory System, Elsevier Biomedical, Amsterdam, 1982, pp. 79–87.
[16] R. Lutfy, How much masking in informational masking?, J. Acoust. Soc. Am. 88 (6) (1990) 2607–2610.
[17] Y. Shimizu, M. Fujiwara, Speech privacy in Japan, relevant feeling, evaluation, in: Proceedings of Inter-Noise 2009, Ottawa, 2009.
[18] Y. Hara, M. Tohyama, K. Miyoshi, Effects of temporal and spectral factors of maskers on speech intelligibility, Appl. Acoust. 73 (9) (2012) 893–899.
[19] K. Saberi, D.R. Perrot, Cognitive restoration of reversed speech, Nature 398 (1999) 760.
[20] A. Kurematsu, K. Takeda, Y. Sagisaka, S. Katagiri, H. Kuwabara, K. Shikano, ATR Japanese speech database as a tool of speech recognition and synthesis, Speech Commun. 9 (1990) 357–363.
[21] D.R. McCloy, P.E. Souza, R.A. Wright, J. Haywood, N. Gehani, S. Rudolph, The PN/NC corpus, version 1.0, http://depts.washington.edu/phonlab/resource/pnnc, 2013 (Last viewed 7/7/2017).
[22] M. Tohyama, T. Koike, Fundamentals of Acoustic Signal Processing, Academic Press, 1998.

CHAPTER

Spectral envelope and source signature analysis

5

CONTENTS

5.1 Response of a linear system to an external source

5.1.1 Spectral product of source signal and impulse response

The convolution for a source signal and the impulse response in the time domain yields the product

$$Y(\omega) = X(\omega)H(\omega) \tag{5.1}$$

Acoustic Signals and Hearing. https://doi.org/10.1016/B978-0-12-816391-7.00013-9

in the frequency domain [1], with the following definitions:

$$\begin{aligned} Y(\omega) &: \text{ Fourier transform of the response } y(t), \\ X(\omega) &: \text{ Fourier transform of the input signal } x(t), \\ H(\omega) &: \text{ similarly for the impulse response } h(t). \end{aligned} \tag{5.2}$$

Suppose that the source (input) signal is composed of several line spectral components or that the signal $x(t)$ is a superposition of sinusoidal functions. The spectral function $Y(\omega)$ for the response to $x(t)$ is composed of the line spectral components at the steady state such that

$$Y(\omega) = \sum_k A_k \delta(\omega - \omega_k) H(\omega), \tag{5.3}$$

where A_k is the complex number with the magnitude and phase of $X(\omega)$ at $\omega = \omega_k$. The source spectral function works as a spectral sampling function for $H(\omega)$. Thus, the continuous spectral function $H(\omega)$ yields the spectral envelope, as in the example shown in Fig. 5.1.

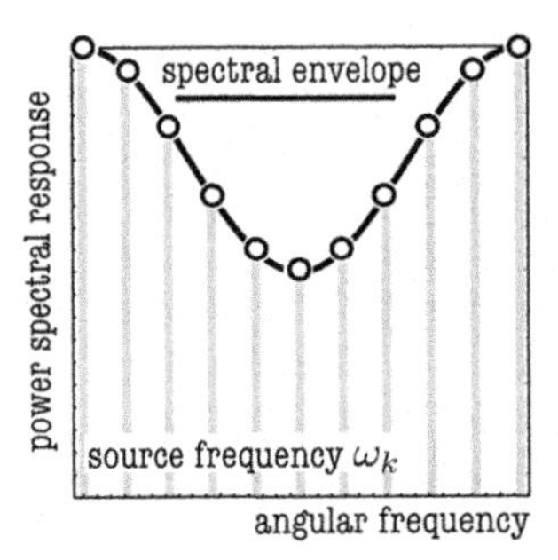

FIGURE 5.1

Power spectral product of line spectral source with a spectral envelope.

The steady-state response $y(t)$ is periodic if the source signal $x(t)$ is periodic, where the source spectral sequence $X(\omega_k)$ is harmonic with the fundamental. This type of response is a model for vowel sounds in the human voice, where $H(\omega)$ represents the spectral function of the vocal tract and $X(\omega_k)$ is the harmonic due to the periodic motion of the vocal cord. This period makes the pitch of the human voice correspond to the fundamental or period of the voice.

In contrast, for a wind instrument like a flute, the spectral function $H(\omega)$ or the Fourier transform of the impulse response $h(t)$ is harmonic with the fundamental and harmonics, while the source signal is mostly broadband random noise. In other words, the spectral function $H(\omega)$ is a superposition of resonant responses (very similar to a line spectral response) for the fundamental and its harmonics. On the other hand, the spectral property of the noise source can be a broadband continuous function. Consequently, the source spectral function is sampled by the line spectral function $H(\omega)$, and the expected power spectral response $|Y(\omega)|^2$ is composed of the fundamental and its harmonics, even if the input source signal is non-periodic and non-harmonic such as random noise. The envelope of the power spectral response represents that for the source signal rather than that for the impulse response.

5.1.2 Frequency characteristics of steady-state vibrations of string

Assume an impulse-like external force or a sinusoidal function of a frequency as the external force. The frequency characteristics for the steady-state response of a vibrating string can be written as a closed form such that

$$H(x_s, x, k) = \mathrm{i}\frac{\sin kx_s \cdot \sin k(L-x)}{\sin kL}, \tag{5.4}$$

where the decaying constant is negligibly small, and x_s and x ($x_s \le x \le L$) are locations for the source and observation points on the string of the length L (m) [1][2][3][4][5].

As described in Section 2.5, the numerator of Eq. (5.4) yields the troughs or the zeros in the complex frequency plane. On the other hand, the denominator makes the poles corresponding to the eigenfrequencies. The poles are independent of the locations of the source and observation points; however, this is not the case for the zeros. The zeros are interlaced with the poles when the location of the observation point is very close to the source: the zeros move to the right (to the higher frequencies) and the number of zeros in the frequency interval of interest decreases as the observation position goes farther from the source location [4][5].

Suppose an external source function $f(t)$ and its Fourier transform $F(\omega)$. The response $y(t)$ and its Fourier transform $Y(\omega)$ are given by

$$y(t) = f * h(t) \quad \text{or} \quad Y(\omega) = F(\omega) \cdot H(\omega) \tag{5.5}$$

following Eq. (5.1), where $h(t)$ denotes the impulse response defined by the inverse Fourier transform of $H(\omega)$. As Eq. (5.4) shows, the frequency characteristic $H(\omega)$ is a series of pulse-like resonance responses including the zeros, under the condition that the decaying constant is negligibly small. In contrast, the external function can be a brief signal in the time domain but with a wide frequency range in the frequency domain. Taking the product of $H(\omega)$ and $F(\omega)$, $H(\omega)$ works for a sampling function of the spectral function of the external force $F(\omega)$. The spectral function of the external force yields the spectral envelope of the response. Fig. 5.2 displays an image of power spectral response that is the product of the spectral function of the external force and frequency characteristics given by Eq. (5.4). The vibration of a string of a musical instrument such as a piano is excited by a hammer that strikes the string. The overall spectral characteristics of string vibration are mostly determined by the spectral envelope (or the outer envelope) made by the striking hammer. The inner envelope due to the zeros depends on the locations of observation and source positions. Fine structures of the vibration represented by line spectral components renders sensation of pitch of the sound in principle.

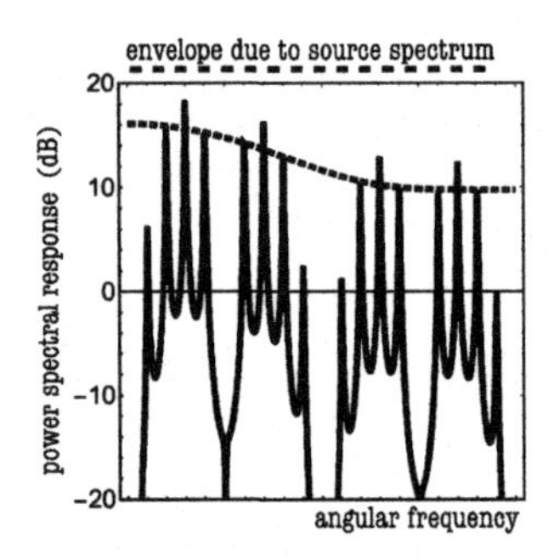

FIGURE 5.2

Power spectral response of a multi-resonance system (e.g., string vibrations, whose frequency characteristics are given by Eq. (5.4)) to a broad frequency band external source (e.g., an impulsive source).

5.1.3 Response to external force in the time and frequency domains

The impulse response can be formulated according to the mirror image method as well as the steady-state spectral characteristics. Suppose the arrangement of the pair of source and observation points on the string shown in Fig. 5.3.

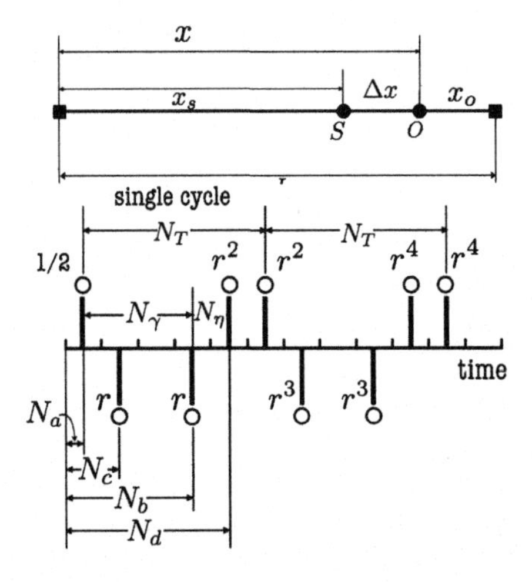

FIGURE 5.3

Locations of source and observation positions on a string of length L (m), and a single cycle of cyclic wave where $(x_s \leq x \leq L)$.

The transfer function that is defined by the z-transform of the impulse response can be written as [2][3].

$$H(z^{-1}) = \sum_n h(n)z^{-n} = \frac{1}{2}z^{-N_a}\frac{(1-|r|z^{-N_{ba}})(1-|r|z^{-N_{ca}})}{1-|r|^2 z^{-N_T}}, \tag{5.6}$$

where

$$N_a T_s = \frac{x - x_s}{c} = \frac{\Delta x}{c} \tag{5.7}$$

$$N_b T_s = \frac{2x_s + \Delta x}{c} \tag{5.8}$$

$$N_c T_s = \frac{\Delta x + 2x_0}{c} \tag{5.9}$$

$$N_d T_s = \frac{2x_s + \Delta x + 2x_0}{c} \tag{5.10}$$

$$N_T T_s = \frac{2x + 2x_0}{c} = \frac{2L}{c} \tag{5.11}$$

$$N_{ba} = N_b - N_a = \frac{2x_s}{cT_s} \tag{5.12}$$

$$N_{ca} = N_c - N_a = \frac{2x_o}{cT_s} \tag{5.13}$$

$$N_{da} = N_d - N_a = \frac{2(x_s + x_o)}{cT_s} = N_{ca} + N_{ba}, \tag{5.14}$$

assuming a negative reflection coefficient $-1 < r < 0$. The numerator of Eq. (5.6) is composed of the direct signal followed by initial reflections that make the cyclic wave. The cyclic wave yields the zeros that depend on the locations of the source and observation positions.

The cyclic wave makes the periodic response that has the poles determined by the denominator or the length of the string independent of the locations for the source and observation positions. Looking at the transfer function given by Eq. (5.6) once more, the "inner" envelope is hidden in the numerator or the cyclic wave independent of the external source function; however, the cyclic wave depends on the arrangement of the pairs of the source and observation positions. Interestingly, the Fourier transform of Eq. (5.6) can be written as

$$H(z^{-1})\Big|_{z=e^{i\Omega}} = i\frac{\sin kx_s \cdot \sin k(L-x)}{\sin kL} \tag{5.15}$$

when $r \to -1$. Eq. (5.15) corresponds to Eq. (5.4), where $k = \omega/c$ (1/m), $\Omega = \omega T_s$, and T_s (s) denotes the sampling period. The numerator indicates the spectral function of a single cycle of the cyclic wave that makes the inner spectral envelope. Fig. 5.4 presents an example of power spectral characteristics with the envelope due to a single cycle of the cyclic wave that depends on the observation and source positions. The power spectral response of the single cycle of the cyclic wave is represented by prominent zeros or the spectral troughs. The total spectral response would be like sampling the frequency characteristics by the poles in the transfer function. The "outer" envelope made by the external force that is represented by a brief signal in the time domain modifies the inner envelope such that

$$\hat{Y}(z^{-1}) \cong F(z^{-1}) \cdot \hat{H}(z^{-1}), \tag{5.16}$$

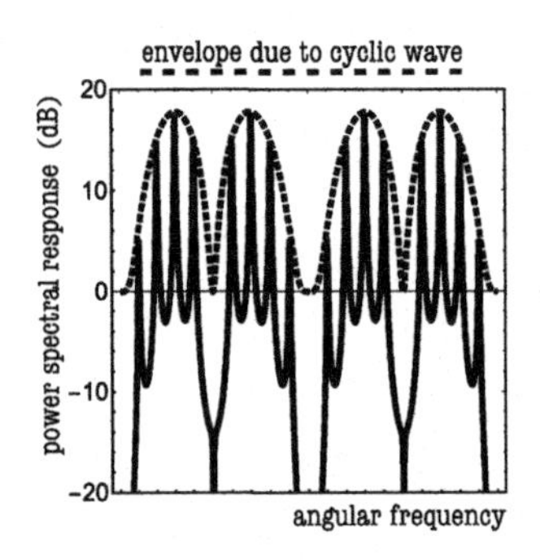

FIGURE 5.4

Example of power spectral characteristics with an envelope due to a single cycle of cyclic wave.

where $\hat{Y}(z^{-1})$ makes the spectral envelope of the response $y(t)$ to the external source function $f(t)$ expressed as Eq. (5.5), and $\hat{H}(z^{-1})$ shows the transfer function for a single cycle of the cyclic wave.

In general the response of the cyclic wave becomes longer due to the external source function; the zeros of the single cycle of the cyclic wave (the inner envelope without the external force) can be preserved, as Eq. (5.16) suggests. On the other hand, the zeros made by the source waveform are independent of the hammering locations for the source. The effects of the hammering function (or the source waveform) at the source locations can be seen in the spectral envelope, which has a wide frequency range in general.

5.2 Clustered line spectral analysis of resonance vibration

5.2.1 Clustered line spectral modeling

A method for estimating a set of clustered line spectral components, in this book called clustered line spectral modeling (CLSM), is formulated according to the least square error (LSE) solution in the frequency domain [2][3][6][7][8]. Exponentially decaying envelopes and slowly changing time envelope of a waveform consist of a number of closely located or clustered sinusoidal components. Beats or amplitude modulation would be an example of the superposition of the paired or clustered sinusoidal components. Fig. 5.5 illustrates an example of beats that are composed of a pair of sinusoidal components. When the observation interval is long enough (panel L) to separate the paired spectral components, the spectral peaks are separately observed corresponding to the pair of frequencies. Consequently spectral peak selection works well [2][3][6][7][8].

In contrast, the observation interval is shorter than the period of the envelope of the beats (panel R), or the observation interval is too short to distinguish the pair of frequencies, the spectral peak no longer indicates the paired components. This occurs

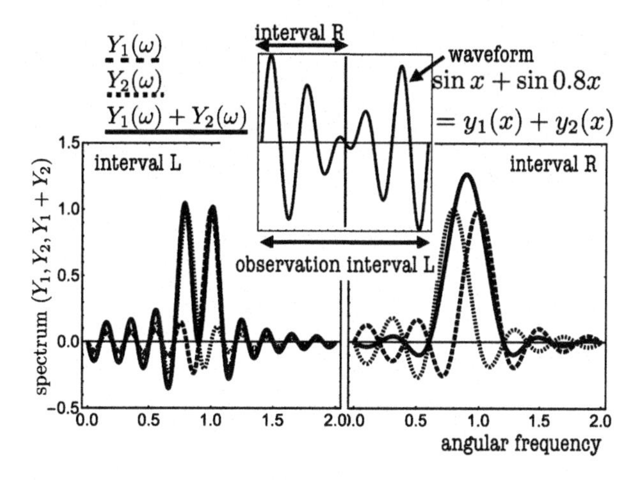

FIGURE 5.5

Separating (panel L) or overlapping (panel R) of power spectral response by observation of beats (center panel) where panel L (R) is spectrum by long enough (too short) interval is applied to superposition of paired sinusoids.

because the spectral function depending on the observation interval is spread over a wider range than the frequency spacing of the paired components. A superposition or overlapping of the spectral functions makes a single central peak instead of separating the corresponding spectral peaks, as displayed in panel R. The CLSM is a method for decomposing the superposed spectral response into corresponding complex functions by solving linear equations based on the LSE criterion on the frequency domain [6]. The method is iteratively repeated by changing conditions of the assuming sinusoids, until the residual error becomes smaller than the desired error range. The details of CLSM are described in the references [2][3][6][7][8][9].

5.2.2 CLSM for piano string vibration

Sound radiated by a piano slowly decays. The vibration of a piano string can be used to demonstrate CLSM for modal resonance vibration [10]. Fig. 5.6 is a waveform (acceleration) to 0.5 (s) for the piano string vibration with its power spectrum that is obtained by the Fourier transform of the entire waveform, and energy decay curve (initial part) of the string vibration. The vibration is excited by a finger from the keyboard. The energy decay curve [11] does not follow an exponential function [12]; however, the vibration waveform reveals a more or less decaying vibration. A number of prominent spectral peaks can be seen in the power spectral record [10]; however, the spacings of the spectral peaks indicate that the vibration is not purely harmonic, although the pitch could be perceived by listening. This type of irregularity or fluctuations in the spacings of the spectral peaks or the non-harmonic nature may represent a property of musical instruments.

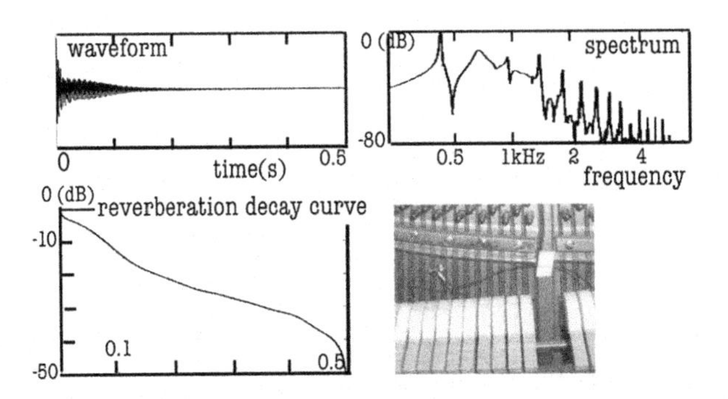

FIGURE 5.6

Waveform up to 0.5 (s) for acceleration of piano string vibration and its power spectrum, and energy decay curve of vibration from Figs. 2 and 4 [10] with a photograph of vibration observation from Fig. 11.10 [3].

The spectral bandwidths or selectivity for the peaks are different corresponding to the harmonics, and the magnitude of the peaks are also different in the frequencies. In particular the difference in the magnitudes of the peaks make the spectral envelope that may indicate the spectral property of the external force. A CLSM approach to the waveform of the string vibration might represent the frequency selectivity around the spectral peaks by using a number of sinusoidal components. The entire spectral envelope might correspond to the spectral characteristics of the modeling error that is still left after CLSM.

The spectral response around the first prominent peak can be represented by five sinusoidal components. Repeating CLSM processing for the seven prominent peaks, the synthesized waveform and its power spectral characteristics are shown in Fig. 5.7 with the original record and the residual error.

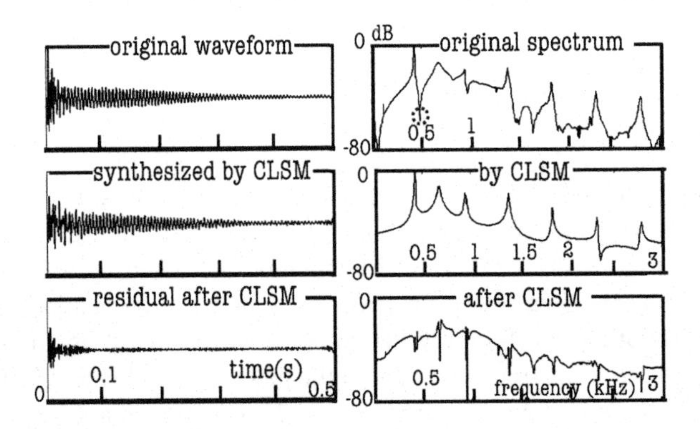

FIGURE 5.7

Waveforms with power spectral records for the original, synthesized by CLSM, and residual error after CLSM from Fig. 6 [10].

Most of spectral peaks are well represented, and the synthesized spectral envelope mostly follows the original envelope. However, the residual error that is defined by the difference between the synthesized and original envelopes can be seen, in particular, in the initial portion of the time waveform. In addition, the representative zero for the original spectral record just after the first prominent peak is missed in the synthesized envelope [10].

The error that is concentrated into the initial portion in the waveform imply the effect of the exciting waveform of the hammer. The observed response may be a superposition of the brief response to the exciting function and the free vibration observed in the entire waveform of the response. The spectral property of the free vibration represented by the resonant responses can be seen in the Fourier transform of the entire response. The example displayed in Fig. 5.7 suggests CLSM may be a possible approach to modal resonance response.

The residual error will be discussed in the next subsection. The representative zero just after the first prominent peak recall the spectral property of a single cyclic waveform as developed in Section 5.1.3. Detailed analysis of the residual after CLSM will reveal the waveform that produces the representative zero [10].

5.2.3 Residual by modeling error and spectral envelope

The envelope of the time waveform with its fine structure can be mostly represented by CLSM as shown in Fig. 5.7. However, the residual error is yielded in the initial portion of the time waveform of the observed record. The power spectral property of the initial portion in the original waveform would be interesting, and motivated us to investigate the residual in detail.

Fig. 5.8 presents the power spectral properties of the original record within the initial 10 (ms), the residual error, and the original entire record (the same as those for the bottom and top of the right column in Fig. 5.7).

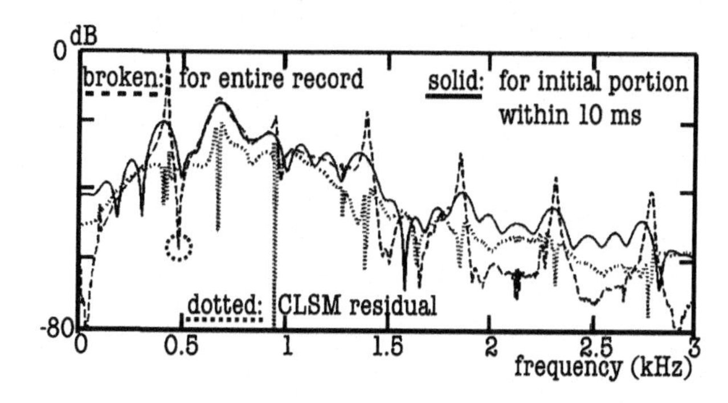

FIGURE 5.8

Power spectral characteristics of the initial part (within 10 (ms)) of the original record, the modeling error, and the original entire response (displayed in Fig. 5.7) from Fig. 7 [10].

Interestingly, the power spectral characteristics of the initial portion follows the spectral envelope of the entire power spectral record. However, the prominent spectral trough in the original record cannot be clearly determined by the power spectral characteristics of the initial portion.

The power spectral behavior of the residual error is similar to that for the initial portion of the original response for the most part. The residual error after CLSM makes the power spectral envelope of the original record, and CLSM also traces the spectral envelope. Fig. 5.8 suggests that CLSM makes a part of the inner spectral envelope. The inner envelope including the prominent trough may be yielded by a superposition of the initial and later parts of the response record of the waveform. A superposition of CLSM and the residual reconstructs the waveform with its entire time-envelope and the inner spectral envelope in the frequency domain according to prominent poles and zeros. The next section analyzes the errors in the time domain. The residual waveform after CLSM would be informative in order to investigate the outer envelope and the prominent zero.

5.3 Clustered time sequence modeling

5.3.1 Representation of brief sequences in the time domain

The CLSM for signal representation in the frequency domain with respect to the dominant spectral peaks uses a number of sinusoidal components that make a time waveform with its slowly varying envelope. In contrast, a brief sequence in the time domain is assumed to be an output of a low-pass filter when a sequence composed of a number of impulses is fed into the filter [10]. Suppose that a sequence could be a superposition of brief sequences, then the sequence could be totally represented as a superposition of the responses of a low-pass filter to a set of pulse trains. The partial pulse trains in the time domain correspond to the sinusoidal components superposed by CLSM for every spectral peak in the frequency domain.

Fig. 5.9 is an image of a short-term waveform decomposing into a pulse train. Take a centered prominent peak in the waveform, as shown in the left panel. The right panel displays time-shifted three-impulse responses of an ideal low-pass filter. Assuming the short-term waveform in the left panel to be a response to the three-pulse train through the low-pass filtering, then the superposition of the time-shifted impulse responses shown in the right panel makes the short-term waveform. The magnitudes and time delays for the pulse train, which is called the clustered time sequence, can be estimated as the LSE solution of a linear equation [10]. The method for estimating the clustered time sequence is called clustered time sequence modeling (CTSM) in this book [8][10].

Fig. 5.10 presents an example of CTSM for a waveform of a brief signal in the time domain. The top row in the right panel shows the original waveform, and the second row illustrates the impulse response record of the low-pass filtering [3][8][10]. Applying the CTSM, the clustered time sequence necessary to represent the prominent peak of the response is obtained according to the LSE criterion, as shown in the

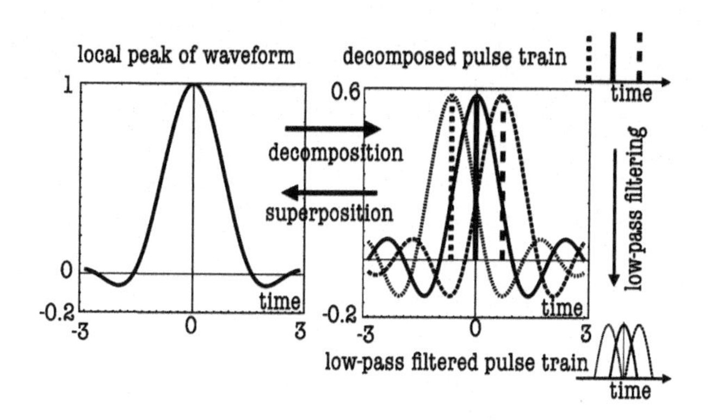

FIGURE 5.9

Image of CTSM analysis for brief response in the time domain.

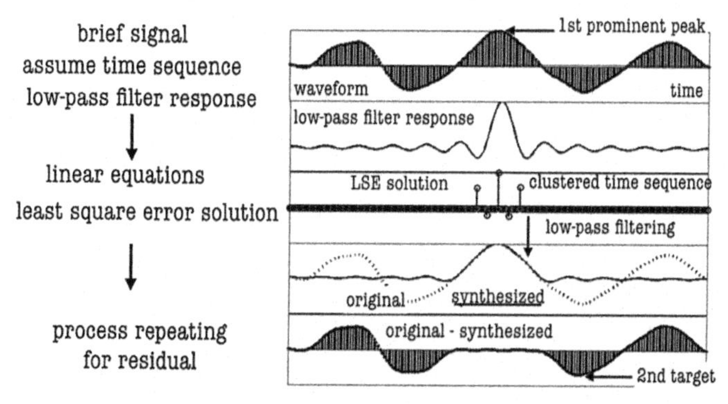

FIGURE 5.10

Example of CTSM for brief signal in time domain from Fig. 10 [10].

third row [1]. The response around the first prominent peak can be fitted by CTSM, as shown in the fourth row. Repeating the CTSM so that the dominant peaks and troughs might not be observed, the residual error might be negligibly small.

5.3.2 Formulation of CTSM by linear equations

The CTSM is a method for representing a waveform around a local peak of a brief signal as a superposition of impulse responses of an ideal low-pass filter [6][10][13]. Suppose that a brief signal $x(n)$ is composed of K impulse responses such that

$$x(n) = \sum_{j=1}^{K} a_j h_j(n), \tag{5.17}$$

which makes a signal peak at $n = n_m$, where $h_j(n)$ denotes the impulse response record for the j-th filter such as an ideal low-pass filter. Assuming the filters are identical or the same impulse response records within the time shifts, then the impulse response for the j-th filter can be written as

$$h_j(n) = h(n - m_j), \tag{5.18}$$

where $h(n)$ represents the identical filters. CTSM gives vector $\hat{\mathbf{a}}$ composed of $\hat{a}_j$, which satisfies the equation

$$x(n) \cong \sum_{j=1}^{P} a_j h_j = \sum_{j=1}^{P} a_j h(n - m_j) \tag{5.19}$$

around the signal peak at $n = n_m$, where $P \leq K$. The time shift for the filter m_j can be written as

$$m_j = n_m - p + j - 1 \quad (1 \leq j \leq P) \tag{5.20}$$

where n_m indicates the location of the targeting peak in the time domain, and

$$p = \begin{cases} \frac{P-1}{2} & P: \text{odd} \\ \frac{P}{2} & P: \text{even.} \end{cases} \tag{5.21}$$

Taking L observation points ($L > P$) in the time sequence, the clustered time sequence $a_j \cong \hat{a}_j$ can be obtained as the LSE solution for L simultaneous equations similar to CLSM; instead, CTSM is formulated in the time domain. The detailed formulation of CTSM (or CLSM) can be seen in reference [6][10].

Taking $L(> P)$ points that define the region in the time domain to represent the targeting signal, the observed signal can be formulated as a vector $\mathbf{x_o}$ composed of $x(n_m - l) \cdots x(n_m - l + L - 1)$, where

$$l = \begin{cases} \frac{L-1}{2} & L: \text{odd} \\ \frac{L}{2} & L: \text{even.} \end{cases} \tag{5.22}$$

The observed impulse responses of the filters in the region make the matrix such that

$$R = \begin{pmatrix} h(n - m_1) & \cdots & h(n - m_P) \\ \vdots & \vdots & \vdots \\ h(n - m_1 + L - 1) & \cdots & h(n - m_P + L - 1) \end{pmatrix}. \tag{5.23}$$

The linear equation

$$R\,\mathbf{a} = \mathbf{x_o} \tag{5.24}$$

holds, where $\mathbf{a}$ is the vector composed of the scalar coefficients a_j, and $\mathbf{x_o}$ denotes the signals $x(n)$ are observed at $n_m - l \leq n \leq n_m - l + L - 1$. The vector $\hat{\mathbf{a}}$ by which the observation signal can be approximately represented as a superposition of the impulse responses can be formulated as the LSE solution such that [1][3][6][8][10]

$$\hat{\mathbf{a}} = \left(R^{\mathrm{T}}R\right)^{-1} R^{\mathrm{T}}\mathbf{x_o}, \tag{5.25}$$

where the signal and impulse responses are assumed to be real numbers. Repeating the CTSM in the time domain so that the prominent signal peaks might disappear, the sequence can be expressed as a set of clustered time-pulse sequences on the basis of the LSE criterion.

5.4 Source waveform estimation from piano string vibration

5.4.1 CTSM of the initial portion of string vibration

The initial portion of the piano string vibration is not well represented, in particular within the initial 10 (ms) following the CLSM. The source information conveyed by a brief sequence in the time domain might be observed by CTSM for the initial portion of the waveform.

Rearranging the synthesized responses for the peaks and troughs in the time order would be intriguing and informative for investigating the source waveform hidden in the vibration. Fig. 5.11 shows synthesized waveforms for the responses of the dominant peaks and troughs in the time order.

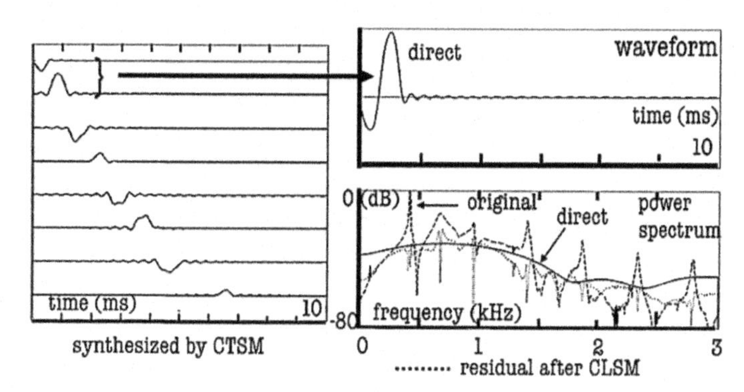

FIGURE 5.11

Left panel: Synthesized responses for dominant peaks and troughs by CTSM in the time order for the initial portion from Fig. 13 [10]. Right panel: Estimated direct waveform (top) and power spectra of direct wave, original waveform, and residual after CLSM from Fig. 14 [10].

The direct wave can be estimated, as shown in Fig. 5.11. The upper and lower panels in the right column give the waveform and its power spectral response. The direct wave yields the power spectral outer envelope for the spectrum of the entire waveform. The power spectral characteristics of the direct wave follows the spectral response of the residual error produced by the CLSM.

The example of CTSM shown in Fig. 5.11 reveals that the initial portion of the response conveys the source information given by the direct wave to the observation point, which yields the power spectral envelope or the outer envelope for the spectral property of the entire waveform. The direct wave with its power spectral property as the outer envelope cannot be accurately estimated following the CLSM, although the residual error possibly contains the source information.

5.4.2 Estimation of single cycle of cyclic wave by CTSM

A cyclic wave that is determined depending on the conditions of the source and observation positions characterizes the spectral envelope and the source waveform, as written in Eq. (5.16). A single cycle of the cyclic wave could be characterized by the zeros, although prominent zeros do not appear in the spectrum of the source waveform shown in Fig. 5.11. The left panel of Fig. 5.12 shows a single cycle of the cyclic wave that is estimated using the results in Fig. 5.11 so that the reflected waves from both of ends of the string might be superposed.

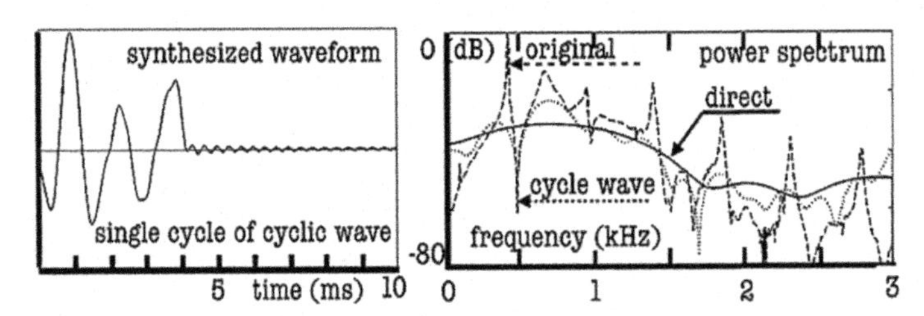

FIGURE 5.12

Estimated single cycle of cyclic wave by CTSM (left panel) with its power spectral response (right panel) from Fig. 15 [10].

The right panel shows the power spectral property of the cyclic wave that traces the first prominent zero just after the first prominent spectral peak including some other zeros for the original spectral characteristics of the string vibration.

Returning to Fig. 5.7, CLSM accurately represents the responses by the poles; however, the prominent zero can be fitted by including the residual error [14]. In contrast, CTSM for the initial portion of the vibration response in the time domain estimates the direct and a single cycle of the cyclic wave that reveals the first prominent zero including some other zeros. The zeros due to the source and observation positions or the exciting source function are inevitable for reconstructing the inner and outer spectral envelopes.

5.5 CLSM and zeros

5.5.1 Poles and zeros

The transfer function is generally written by the z-transform of the impulse response in terms of the poles and zeros. Suppose that the transfer function could be formulated such that [4][5][14]

$$H(z^{-1}) = \frac{A}{1 - az^{-1}} + \frac{B}{1 - bz^{-1}} + R(z^{-1}), \tag{5.26}$$

which represents a sum of the prominent pole responses and superposition of other responses determined by other poles. Here A and B are the residues of the dominant poles, and $R(z^{-1})$ shows the remainder responses by other poles [4][5]. The zeros are the roots that satisfy the equation

$$H(z^{-1})\Big|_{z=z_0} = 0. \tag{5.27}$$

The zeros depend on the sign change in A and B [4][5]. A zero can be found between the prominent poles subject to that A and B are the same sign. On the other hand, the locations of the zeros depend on the remainder function for a pair of different-sign residues [4][5][15][16].

The CLSM represents the dominant pole responses by superposition of a clustered sinusoidal components. An interesting question would be whether the zero between the dominant poles can be simultaneously represented by the CLSM or not.

5.5.2 CLSM for response due to a pair of poles with same-sign residues

Return to the transfer function in Eq. (5.26) assuming $R(z^{-1}) \cong 0$ for the same-sign residues $A = B = 1$. The zeros of the transfer function can be found between the poles. The left panel of Fig. 5.13 shows an example of the transfer function with a pair of same-sign residues to which CLSM approach is taken.

The upper graphs in the left panel show the power spectral responses for the original (solid) and synthesized (dotted) responses by CLSM. The lower drawing illustrates the phase of the residual error after CLSM. The synthesized response in the upper graph follows mostly the original (or observed) response even around the trough due to the zero. Actually, the phase response in the lower panel shows the opposite π-phase change to those for the poles, so that the residual error might be almost zero. The trough (zero) can be well fitted by CLSM, as can the dominant poles [14].

5.5.3 Zeros of response to excitation source

A zero of the transfer function with the same-sign residues might be represented by CLSM [14], as shown in the left panel of Fig. 5.13. The response to which CLSM is

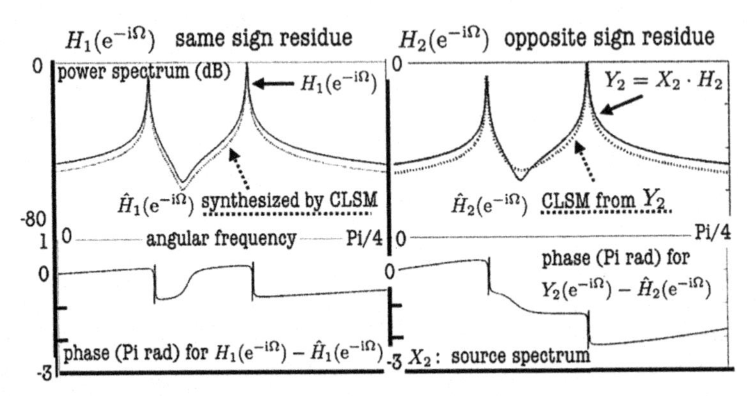

FIGURE 5.13

Examples of CLSM for transfer functions. Left panel: For H_1 composed of a pair of poles inside the unit circle with same-sign residues, Right panel: For response to a source function, such as $Y_2 = X_2 \cdot H_2$ through a transfer function with opposite-sign residues from Fig. 1 [14].

applied would not be the transfer function but could be a response to a source signal through the transfer function in general. Suppose a source signal $x(n)$ and the impulse response of the system of interest $h(n)$. The output response $y(n)$ can be expressed as the convolution such that

$$y(n) = x * h(n) \quad \text{or} \quad Y(z^{-1}) = X(z^{-1})H(z^{-1}), \tag{5.28}$$

where $Y(z^{-1}), X(z^{-1})$, and $H(z^{-1})$ are the z-transforms of $y(n), x(n)$, and $h(n)$, respectively. The z-transform of $x(n)$ is defined as

$$X(z^{-1}) = \sum_{n=0}^{N-1} x(n)z^{-n}, \tag{5.29}$$

where $x(n)$ is defined in the interval $0 \le n \le N-1$. Substituting Eq. (5.29) for Eq. (5.28),

$$Y(z^{-1}) = H(z^{-1}) + \left[\sum_{n=1}^{N-1} x(n)z^{-n}\right] H(z^{-1}) = H(z^{-1}) + R_X(z^{-1}) \tag{5.30}$$

is obtained subject to $x(0) = 1$ [14], where

$$R_X(z^{-1}) = \left[\sum_{n=1}^{N-1} x(n)z^{-n}\right] H(z^{-1}). \tag{5.31}$$

Comparing Eq. (5.30) with Eq. (5.26), $R_X(z^{-1})$ can be understood as the remainder function, under the condition that the transfer function can be expressed as a superposition of the responses due to a pair of the poles. The formulation of Eq. (5.30) implies that the source function may change the zeros under the opposite residue-sign condition for the pair of poles because the zeros are sensitive to the remainder function when the residues have opposite signs [4][5][15][16].

5.5.4 Poles, zeros, and accumulated phase of response including excitation source

Take again an example of responses due to a pair of poles with same-sign residues, such as [4][5]

$$H_1(z^{-1}) = \frac{1}{1 - az^{-1}} + \frac{1}{1 - bz^{-1}} = \frac{z(z - z_0)}{(z - a)(z - b)}, \tag{5.32}$$

where

$$z_0 = \frac{a + b}{2}, \tag{5.33}$$

a and b denote complex numbers for the pair of poles, and the poles and zeros are inside the unit circle in the complex frequency plane. Namely, Eq. (5.32) has two poles at $z = a$ and $z = b$, and two zeros at $z = 0, (a + b)/2$ within the unit circle where $|z| \leq 1$. No phase accumulation is defined by

$$\theta(\Omega)|_{\Omega=2\pi} - \theta(\Omega)|_{\Omega=0} = \Phi(2\pi) = 0, \tag{5.34}$$

where θ is the phase angle when $z = e^{i\Omega}$. The phase accumulation by two poles can be canceled by that for the zeros. This phase cancelation explains why there is no phase accumulation.

In contrast, suppose the transfer function is written as [4][5]

$$H_2(z^{-1}) = \frac{1}{1 - az^{-1}} + \frac{-1}{1 - bz^{-1}} = \frac{z(a - b)}{(z - a)(z - b)}. \tag{5.35}$$

Similarly to Eq. (5.32), Eq. (5.35) has two poles at $z = a$ and $z = b$, but only a single zero $z_0 = 0$ for $|z| \leq 1$. The accumulated phase becomes

$$\Phi(2\pi) = -2\pi \tag{5.36}$$

because the accumulated phase due to the pair of poles cannot be canceled by that for the single zero.

No zeros are located between the poles for the opposite sign residues if no remainder function is available. However, the zeros are sensitive to the remainder function [4][5][15][16]. A source function exciting the system of interest can be interpreted as the remainder function, as suggested by Eq. (5.30) [14]. Suppose an example of a source function

$$X_2(z^{-1}) = 1 - cz^{-1}, \tag{5.37}$$

where c gives the zero for the source function. The response to the source through the transfer function $H_2(z^{-1})$ is given by

$$Y_2(z^{-1}) = X_2(z^{-1})H_2(z^{-1}) = \frac{(z-c)(a-b)}{(z-a)(z-b)}, \tag{5.38}$$

where $X_2(z^{-1})$ is independent of the transfer function $H_2(z^{-1})$.

Eq. (5.38) has a zero at $z = c$ in addition to the pair of poles, even under opposite sign residues, because the zero $z = c$ is due to the source function. The accumulated phase of the response depends on the location of the zero ($z_0 = c$). If the zero is located inside the unit circle ($|z_0| \leq 1$) the accumulated phase is -2π because of the cancelation of the zero and one of the poles; however, the zero located outside the unit circle ($|z_0| > 1$) of the accumulated phase becomes -4π because there is no cancelation between the zero and any of the poles [1][2][3][4][5][8][14][15][16]. An interesting question would be whether the zero due to the source function can be synthesized by CLSM for the dominant poles.

5.5.5 CLSM for responses including source function

Recall the left panel of Fig. 5.13 once more where CLSM fits the zero as well as the pair of poles with same-sign residues. In contrast, the right panel of the same figure presents another CLSM example for the response given by Eq. (5.38) that includes the source function $X(z^{-1})$, where the transfer function $H_2(z^{-1})$ is written as Eq. (5.35) [14]. The upper panel in the right column gives the observed and synthesized (corresponding to $\hat{H}_2$) power spectral responses. An interesting point would be the residual error around the trough between the poles. The error cannot be evaluated by the magnitude of the error around the trough at which the magnitude might be sufficiently small. In contrast, the phase response of the residual (error) in the lower graph seems not to be the phase behavior around the zero. The trough or zero due to the source condition independent of the transfer function cannot be well represented by CLSM for the poles of the transfer function. The residual after CLSM may be necessary in order to represent the zeros due to the source signature.

Fig. 5.14 reorganizes the original and synthesized power spectral records, as in Figs. 5.7 and 5.8, where the lower panel is a close-up around the prominent zero just after the first prominent spectral peak [14]. In the close-up a comparison of the solid and dotted lines would be interest. The solid line (CLSM) is a superposition of the synthesized responses only for the pair of poles adjacent to the trough, while the dotted line (remainder: Rm) shows the remainder function [4][5] in Eq. (5.26), which gives a sum of the contribution from other poles outside the corresponding frequency bands. However, the dotted line (remainder) may be too small to create the zero by canceling the solid line (CLSM) at the trough.

On the other hand, the dash-dotted line gives the remaining error after CLSM fitting. The error is needed here to create the prominent trough made by the zero as a result of canceling. The error function must be composed of the sum of the

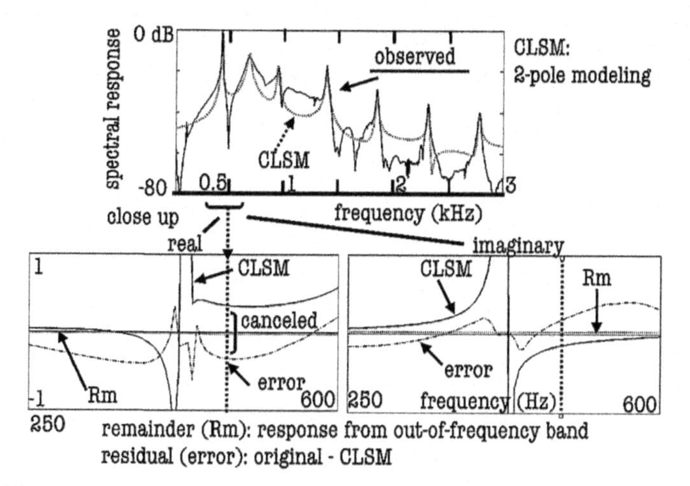

FIGURE 5.14

CLSM of piano string vibration, similar to Fig. 5.7 and Fig. 2 [14].

out-of-band response, the source function in this example, and the total "remainder" function may create the zero between the pair of poles with opposite sign of residues.

The cyclic wave including the source function makes some other zeros around 1500 (Hz), as shown in Fig. 5.12. In addition the remainder function excluding the source effect can yield the double-zero around 2.5 (kHz) that appears in Fig. 5.14 where the contribution from the other poles lower than 2.5(kHz) might be needed. Double-zeros are examples of zeros under opposite-sign residues caused by the remainder function [4][5][15][16].

Zeros would be identified following a different scheme from CLSM for the poles, however. Another approach to the identification of zeros will be developed in the next section.

5.6 Identification of zeros for source signature analysis

5.6.1 Identification of zeros using linear equations

A source signature such as a striking hammer on a piano string can be modeled by a brief train of pulses in the time domain [17]. The z-transform of a time sequence can be expressed as

$$H(z^{-1}) = \sum_{n=0}^{N-1} h(n)z^{-n} \tag{5.39}$$

$$= K\prod_{k=1}^{N-1}(1 - a_k z^{-1}) = Kz^{-(N-1)}\prod_{k=1}^{N-1}(z - a_k),$$

where $h(n)$ denotes the time sequence representing a response in the time domain, and K is a constant. The zeros of $H(z^{-1})$ are determined by the roots of the polynomial. A single zero, for example

$$1 - az_0^{-1} = 0, \tag{5.40}$$

is specified by a superposition of the direct pulse and a unit-sample delayed pulse that makes the combination of adjacent-pairing time pulse, when K is normalized to unity. The vector $(1, a)^{\mathrm{T}}$ gives the representation of the response around the spectral trough by z_0 in the time domain, where a is a complex number that corresponds to a single zero.

Suppose that an unknown vector $\mathbf{a} = (1, a)^{\mathrm{T}}$ satisfies the equation

$$A\mathbf{a} = \mathbf{b}, \tag{5.41}$$

where $\mathbf{b}$ shows the vector composed of the observation records in complex functions around the spectral trough, A is the matrix

$$A = \begin{pmatrix} 1 & \mathrm{e}^{-\mathrm{i}\Omega_0} \\ \vdots & \vdots \\ 1 & \mathrm{e}^{-\mathrm{i}\Omega_{L-1}} \end{pmatrix}, \tag{5.42}$$

and L denotes the number of entries of the vector $\mathbf{b}$. According to the LSE criterion (or the LSE solution), the solution $\hat{\mathbf{a}}$ is given by

$$\hat{\mathbf{a}} = (A^{\mathrm{T}}A)^{-1}A^{\mathrm{T}}\mathbf{b}. \tag{5.43}$$

The representation for a spectral trough is called adjacent paring time-pulse modeling (APTM) in this book [1][3][8][17].

The CLSM is based on the LSE solution around the pole, while the APTM is based on that around the spectral trough. In contrast to CLSM, which gives a prominent spectral peak in the frequency domain, APTM represents the time sequence that yields a spectral trough.

5.6.2 Spectral trough selection

The procedure for APTM can be iteratively repeated by subtraction in the logarithmic scale, for example

$$H_0(z^{-1}) = H_1(z^{-1})H_2(z^{-1}), \tag{5.44}$$

where $H_1(z^{-1})$ and $H_2(z^{-1})$ have a single zero and a trough, respectively. After estimating $H_1(z^{-1})$ by $\hat{H}_1(z^{-1})$, subtraction such that

$$\log_{\mathrm{e}} H_0(z^{-1}) - \log_{\mathrm{e}} \hat{H}_1(z^{-1}) \cong \log_{\mathrm{e}} H_2(z^{-1}), \tag{5.45}$$

makes mostly $H_2(z^{-1})$, which can be the second target for modeling the second zero (or spectral trough). Spectral smoothing can be performed, without spectral equalization or filtering, by using the spectral trough selection according to APTM [8][17].

Prominent spectral troughs are related to the source signal signature, as described in the previous example of the piano string vibration. Fig. 5.15 illustrates an example of the spectral trough selection for the piano string vibration described in the previous section [17].

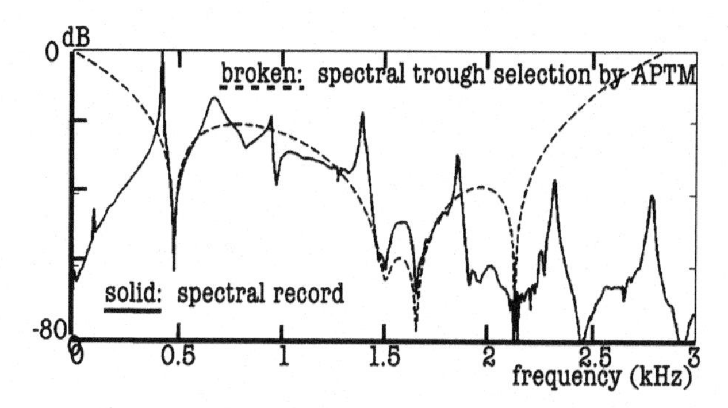

FIGURE 5.15

APTM for spectral trough selection of piano string vibration from Fig. 4 [17].

The first three troughs can be produced by a single cycle of the cyclic wave including the source function. The APTM for spectral trough selection works well for the first three prominent troughs, and the next trough can also be represented by repeating the approach. Modeling responses to a source function through the transfer function such as CLSM, CTSM, and APTM dealt with in this chapter would be possible ways to identify the transfer function including the source signatures in both of the time and frequency domains [6][10][14][17].

5.7 Exercises

1. Confirm Eq. (5.5).

2. Derive Eqs. (5.6) and (5.15).

3. Derive Eq. (5.25).

4. Suppose a matrix A and $A\mathbf{x} = \mathbf{b}$. Explain the difference between the operations $A^{-1}A$ and $A^{\mathrm{T}}A$.

5. Show the poles and zeros of Eq. (5.6).

6. Explain terminologies listed below:

(1) convolution (2) impulse response (3) spectral product
(4) spectral envelope (5) pole (6) zero
(7) cyclic wave (8) transfer function (9) inner spectral envelope
(10) outer spectral envelope (11) spectral overlap (12) least square error criterion
(13) direct wave (14) mirror image method.

References

[1] M. Tohyama, T. Koike, Fundamentals of Acoustic Signal Processing, Academic Press, 1998.
[2] M. Tohyama, Sound in the Time Domain, Springer, 2017.
[3] M. Tohyama, Sound and Signals, Springer, 2011.
[4] R.H. Lyon, Progressive phase trends in multi-degree-of-freedom systems, J. Acoust. Soc. Am. 73 (4) (1983) 1223–1228.
[5] R.H. Lyon, Range and frequency dependence of transfer function phase, J. Acoust. Soc. Am. 76 (5) (1984) 1435–1437.
[6] M. Kazama, K. Yoshida, M. Tohyama, Signal representation including waveform envelope by clustered line-spectrum modeling, J. Audio Eng. Soc. 51 (3) (2003) 123–137.
[7] O. Yasojima, Y. Takahashi, M. Tohyama, Resonant bandwidth estimation of vowels using clustered-line spectrum modeling for pressure speech waveforms, in: Int. Symp. Signal Processing and Information Technology, IEEE, 2006, pp. 589–593.
[8] M. Tohyama, Waveform Analysis of Sound, Springer, 2015.
[9] T. Taniguchi, M. Tohyama, K. Shirai, Detection of speech and music based on spectral tracking, Speech Commun. 50 (2008) 547–563.
[10] T. Hasegawa, M. Tohyama, Analysis of spectral and temporal waveforms of piano-string vibration, J. Audio Eng. Soc. 60 (4) (2012) 237–245.
[11] M.R. Schroeder, New method of measuring of reverberation time, J. Acoust. Soc. Am. 37 (3) (1965) 409–412.
[12] G. Weinreich, Coupled piano strings, J. Acoust. Soc. Am. 62 (6) (1977) 1474–1484.
[13] H. Nakajima, M. Tanaka, M. Tohyama, Signal representation and inverse filtering using recursive vector projection (in Japanese with English abstract), J. Inst. Electron. Inf. Comm. Eng. Jpn. J 83-A (4) (2000) 353–360.
[14] T. Hasegawa, M. Tohyama, Separation of zeros for source signature identification under reverberant path condition, J. Acoust. Soc. Am. 130 (4) (2011) EL271–EL275.
[15] M. Tohyama, R.H. Lyon, Zeros of a transfer function in a multi-degree-of-freedom system, J. Acoust. Soc. Am. 86 (5) (1989) 1854–1863.
[16] M. Tohyama, R.H. Lyon, T. Koike, Reverberant phase in a room and zeros in the complex frequency plane, J. Acoust. Soc. Am. 89 (4) (1991) 1701–1707.
[17] T. Hasegawa, M. Tohyama, Source signature identification by using pole/zero modeling of transfer function, in: Inter Noise 2011, 2011, 431616.

CHAPTER

Room reverberation theory and transfer function

6

CONTENTS

6.1 Reverberation time formulas

6.1.1 Density of reflection sounds

Reverberation is a process of collisions of sound into the surrounding walls (or reflections from the walls). The sound field composed of the direct and reflection sounds can be basically represented by an array of mirror-image sources. Suppose a rectangular room surrounded by hard walls where a sound source is located at the center. The mirror-image sources can be displayed as shown in Fig. 6.1.

Acoustic Signals and Hearing. https://doi.org/10.1016/B978-0-12-816391-7.00014-0

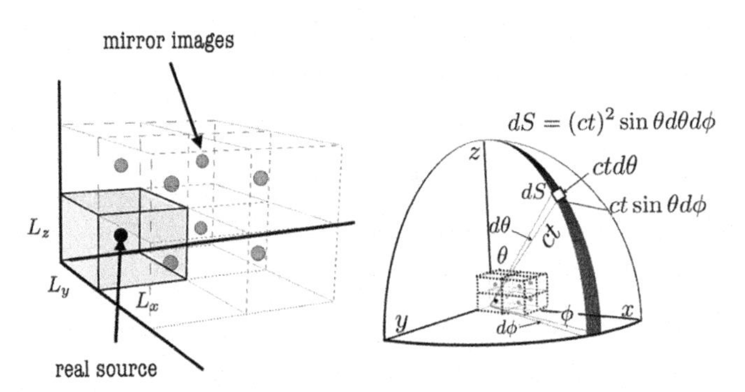

FIGURE 6.1

Example of mirror-image sources and the spherical coordinates in which image sources are located.

The number of reflected sounds which come back to the source position (the center of the room) within t seconds after the sound source radiates an ideal impulse can be estimated by [1][2]

$$N(t) \cong \frac{4\pi(ct)^3}{3V}, \tag{6.1}$$

where V (m^3) denotes the volume of the room and c (m/s) gives the speed of sound in the room. According to the estimate in Eq. (6.1) the density of the reflection can be defined as

$$n(t) = \frac{d}{dt}N(t) \cong \frac{4\pi c^3 t^2}{V}, \qquad (1/\mathrm{s}) \tag{6.2}$$

which deceases in inverse proportion to the room volume, and increases in proportion to t^2. The inverse proportion to the room volume, but proportion to the square of time is the typical nature of the reflected sounds during propagation in a three-dimensional space.

6.1.2 Number of collisions into surrounding walls

Consider a reflected sound that comes back to the sound source at the center of the room within t seconds after the sound source radiates an ideal impulse. Take a mirror-image source at a position (θ, ϕ, ct) in the coordinate system shown in the right panel of Fig. 6.1. The number of collisions at the $x-$wall perpendicular to the $x-$axis can be estimated by

$$R_x(\theta, \phi, t) = \frac{ct \sin\theta \cos\phi}{L_x} \tag{6.3}$$

for the reflected sound coming back from the mirror-image source located at $\theta, \phi, r = ct$. Similarly, the number of collisions for the $y-$ and $z-$walls are

$$R_y(\theta, \phi, t) = \frac{ct \sin\theta \sin\phi}{L_y} \quad \text{and} \quad R_z(\theta, \phi, t) = \frac{ct \cos\theta}{L_z}, \tag{6.4}$$

respectively, where L_x, L_y, L_z (m) denote the lengths of the sides. The number of collisions depends on the locations of mirror-image sources, and is given by [1][2][3]

$$R(\theta, \phi, t) = R_x(\theta, \phi, t) + R_y(\theta, \phi, t) + R_z(\theta, \phi, t). \tag{6.5}$$

6.1.3 Collisions and mean free path

Take the ensemble average of the numbers of the collisions for the whole of mirror-image sources. Define the density of mirror-image sources such that

$$p(\theta, \phi) = \frac{dS}{4\pi(ct)^2/8} = \frac{\sin\theta d\phi d\theta}{\pi/2}, \tag{6.6}$$

assuming that the mirror-image sources are densely distributed [1][2][3]. The average for R_x becomes

$$\mathrm{E}[R_x] = \frac{2}{\pi}\int_0^{\pi/2} d\phi \int_0^{\pi/2} R_x \sin\theta d\theta = \frac{ct}{2L_x}. \tag{6.7}$$

Similarly,

$$\mathrm{E}[R_y] = \frac{ct}{2L_y} \quad \text{and} \quad \mathrm{E}[R_z] = \frac{ct}{2L_z} \tag{6.8}$$

are obtained, respectively. The ensemble average of the total number of collisions is given by

$$\overline{R} = \mathrm{E}[R] = \mathrm{E}[R_x] + \mathrm{E}[R_y] + \mathrm{E}[R_z] = \frac{S}{4V}ct = \frac{ct}{MFP}, \tag{6.9}$$

where

$$V = L_x L_y L_z \ (\mathrm{m}^3), \quad S = 2(L_x L_y + L_y L_z + L_z L_x) \ (\mathrm{m}^2), \tag{6.10}$$

$$MFP = \frac{4V}{S} \ (\mathrm{m}), \tag{6.11}$$

and MFP is called the mean free path of the sound propagation in the room, and gives an estimate of the averaged length of sound paths of inter-collisions in the sound space.

6.1.4 Reverberation energy decay and reverberation time

The sound intensity at the source position is a function of time, and is composed of the intensity due to the reflected sounds coming from the mirror-image sources. Suppose that the sound power output of the source is P_0(W). The expected sound intensity (or the sound power flow across the unit area) due to the mirror-image sources at (θ, ϕ, t) is formulated by

$$I(t) = \frac{P_0(1-\alpha)^{\overline{R}(t)}}{4\pi(ct)^2} n(t)dt \qquad (\mathrm{W/m^2}) \tag{6.12}$$

assuming symmetric spherical waves for the reflected sounds, where α $(0 < \alpha < 1)$ denotes the averaged absorption coefficient of surrounding walls, $\overline{R}(t)$ gives the expectation of the number of collisions for a reflected sound from the mirror-image source at (θ, ϕ, t). Introducing Eq. (6.2),

$$I(t) = I_0(1-\alpha)^{\overline{R}(t)} \text{ and } I_0 = \frac{cdt}{V} P_0 \ (\mathrm{W/m^2}), \tag{6.13}$$

where the decrease by $(ct)^2$ in intensity following the denominator of Eq. (6.12) is compensated by an increase in the density of the mirror-image sources $n(t)$. This compensation is a typical phenomenon in three-dimensional space.

Taking the natural logarithm of the intensity such that

$$\log \frac{I(t)}{I_0} = \overline{R}(t) \log(1-\alpha), \tag{6.14}$$

then Eq. (6.14) becomes

$$\overline{R}(t) \log(1-\alpha) = \frac{cSt}{4V} \log(1-\alpha) \tag{6.15}$$

according to Eq. (6.9). The reverberation time is defined as the time interval when the intensity I decreases to 10^{-6} of I_0 at the initial state. By setting

$$\frac{cS}{4V} T_R \log(1-\alpha) = \log 10^{-6}, \tag{6.16}$$

the reverberation time T_R can be formulated as

$$T_R = \frac{4 \cdot 6 \cdot \log 10}{c} \cdot \frac{V}{-\log(1-\alpha)S} \cong \frac{4}{c} \cdot 13.82 \cdot \frac{V}{A} \cong 0.163 \frac{V}{A}, \qquad (\mathrm{s}) \tag{6.17}$$

where

$$A = -\log(1-\alpha)S \qquad (\mathrm{m^2}), \tag{6.18}$$

which is called the equivalent absorption area, and $c \cong 340$ (m/s) are obtained.

According to Eq. (6.13), the average of the intensity I_s at the steady state before reverberation (or the sound source stops) can be expressed as

$$I_s = \int_0^\infty \frac{P_0(1-\alpha)^{\overline{R}c}}{V} dt = \frac{4P_0}{-\log(1-\alpha)S}. \qquad (\mathrm{W/m^2}) \tag{6.19}$$

The sound intensity in a room increases as the sound power output of the source increases, while it decreases as the sound absorption of the walls increases. The sound intensity decay curve, on average, can be formulated as

$$I(t) = \int_t^{\infty} \frac{P_0(1-\alpha)^{\mathrm{E}[R]}c}{V} dt' = \frac{4P_0}{-\log(1-\alpha)S} \mathrm{e}^{\frac{cS}{4V} t \log(1-\alpha)}. \quad (\mathrm{W/m^2}) \qquad (6.20)$$

6.2 Reverberation response to direct sound

6.2.1 Reverberation decay curve for a linear system

Reverberation energy decay curves can be formulated for linear systems by using impulse responses [4]. Suppose a source of white noise $x(t)$ that is an input signal fed into a linear system stops at $t = 0$. Taking the impulse response $h(t)$ from the source position to a receiving position, the response $y(t)$ to the white noise signal $x(t)$ can be expressed as

$$y(t) = x * h(t), \qquad (6.21)$$

where $*$ indicates convolution. According to the random nature of white noise $n(t)$, the ensemble average of $y^2(t)$ for $t \geq 0$ becomes [4]

$$\begin{aligned} \mathrm{E}[y^2(t)] &= \mathrm{E}\left[\int_t^{\infty} h(\tau_1)x(t-\tau_1)d\tau_1 \int_t^{\infty} h(\tau_2)x(t-\tau_2)d\tau_2\right] \\ &= N^2 \int_t^{\infty} h^2(\tau)d\tau, \end{aligned} \qquad (6.22)$$

which yields the reverberation decay curve after the white noise stopped, where

$$\mathrm{E}[x(t-\tau_1)x(t-\tau_2)] = N^2\delta(\tau_2 - \tau_1) \qquad (6.23)$$

because of the random nature of white noise, and

$$x(t) = \begin{cases} x(t) & t \leq 0 \\ 0 & t > 0. \end{cases} \qquad (6.24)$$

Suppose the impulse response is

$$h(t) = \mathrm{e}^{-\delta t}\mathrm{e}^{i\omega t} \qquad (6.25)$$

in a complex function for a single resonator. The reverberation energy decay curve is given by

$$\int_t^{\infty} |h^2(\tau)|d\tau = \int_t^{\infty} \mathrm{e}^{-2\delta\tau}d\tau = \frac{1}{2\delta}\mathrm{e}^{-2\delta t} \to 0, \qquad (t \to \infty) \qquad (6.26)$$

which indicates the squared envelope of the impulse response within a scaling factor.

6.2.2 Buildup and decay for reverberant space

Suppose the noise source $x(t)$ starts at $t = 0$. The response to the noise source after the source started is formulated as

$$y(t) = x * h(t) = \int_0^t h(\tau)x(t-\tau)d\tau. \tag{6.27}$$

The ensemble average of the squared response becomes

$$u^2(t) = \mathrm{E}[y^2(t)] = N^2 \int_0^t h^2(\tau)d\tau, \tag{6.28}$$

which indicates the buildup process from the starting to the steady state of the response.

Take again the example of the impulse response from Eq. (6.25). The buildup response becomes

$$u^2(t) = \frac{1}{2\delta}\left(1 - \mathrm{e}^{-2\delta t}\right) \to \frac{1}{2\delta}. \qquad (t \to \infty) \tag{6.29}$$

A buildup process reaches the steady-state response to be constant as t passes long time assuming an appropriate damping condition.

Compare the three kinds of responses:

$$u^2(t) = \begin{cases} \frac{1}{2\delta}\left(1 - \mathrm{e}^{-2\delta t}\right) & \text{buildup} \\ \frac{1}{2\delta} & \text{steady state} \\ \frac{1}{2\delta}\mathrm{e}^{-2\delta t}. & \text{decay process} \end{cases} \tag{6.30}$$

The response at the steady state becomes high as the damping factor (or decay rate 2δ) decreases. On the other hand, the transient response quickly builds up to reach the steady state as the decay rate goes high. Reverberation becomes rich in a listening room as the decay rate decreases; however, sound localization might be unclear because of the slow response for the transient state.

Fig. 6.2 is an image of the buildup and decay responses following Eq. (6.30). The absorbed power in the decaying process is expressed as

$$b^2(t) = \frac{1}{2\delta} - \frac{1}{2\delta}\mathrm{e}^{-2\delta t} = \frac{1}{2\delta}\left(1 - \mathrm{e}^{-2\delta t}\right), \tag{6.31}$$

which, interestingly, is equal to the response in the buildup process. Recalling the mirror-image sources that create the reverberation process (or decaying process), the buildup process is a superposition of the direct and reflected sounds. The buildup and reverberation processes are complementary of processes to each other; the direct one is followed by the reflections of switching on (or off) mirror-image sources.

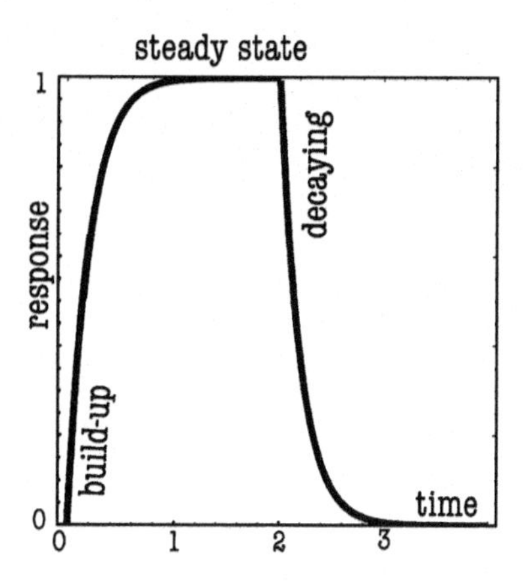

FIGURE 6.2

Examples of buildup and decay responses.

Taking the first derivative of the buildup response with respect to the time, the derivative reaches zero at the steady state. Sound power supplied by the source and absorbed power by the surrounding walls are balanced at the steady state so that the response might be steady.

The initial decay rate for the decaying process is

$$-\frac{d}{dt}U_s e^{-2\delta t}\bigg|_{t=0} = 2\delta U_s, \qquad \text{(W)} \tag{6.32}$$

where $U_s = I_s V/c$ (J) is the energy response at the steady state [5]. The initial decay rate must be equal to the source power such that

$$2\delta U_s = -\frac{d}{dt}\left[\frac{V}{c}I(t)\right]\bigg|_{t=0} = P_0 \qquad \text{(W)} \tag{6.33}$$

following Eq. (6.20). The relation holds at the steady state in which the first derivative of the energy response is zero. According to Eq. (6.19)

$$2\delta = \frac{-\log(1-\alpha)cS}{4V}. \quad (1/\text{s}) \tag{6.34}$$

The sound power output of a sound source is an averaged behavior at the steady-state response of a source. The initial decay rate given by Eq. (6.33), however, implies that the sound power output of a source might also be related to the transient response of the source in a reverberant field.

6.3 Room transfer function

6.3.1 Frequency characteristics and impulse response of sound field in rectangular room

Suppose a point source (a limit when the radius a (m) $\rightarrow 0$ for a spherical source) radiates a sinusoidal wave in a rectangular room surrounded by a rigid wall. A response observed at a receiving point makes the frequency characteristics for the room at the steady state. According to the linear system theory, the response observed at a receiving position is formulated by the convolution such that

$$y(t) = x * h(t), \qquad (6.35)$$

where $x(t)$ denotes the signal radiated by the source and $h(t)$ gives the impulse response between the source and receiving positions.

Following the mirror-image theory, the impulse response can be written as [1][2][6][7]

$$h(\mathbf{x}_s, \mathbf{x}_o, t) = \hat{Q} \sum_{P=1}^{8} \sum_{N=-\infty}^{\infty} \frac{8}{\Lambda_N} \frac{\delta(t - |\mathbf{R}_P + \mathbf{R}_N|/c)}{4\pi |\mathbf{R}_P + \mathbf{R}_N|}, \quad (\mathrm{m}^2/\mathrm{s}) \qquad (6.36)$$

where

$$\hat{Q} = Q/\Delta\omega \quad (\mathrm{m}^3) \text{ and } Q: \text{ volume velocity of the source} \quad (\mathrm{m}^3/\mathrm{s}) \qquad (6.37)$$

$$\Lambda_N: \text{ normalize coefficients for image sources such that} \qquad (6.38)$$

$$= \begin{cases} 8 & l \neq 0, m \neq 0, n \neq 0 \\ 4 & \text{one of } (l, m, n) \text{ is } 0 \\ 2 & \text{two of } (l, m, n) \text{ are } 0 \end{cases} \qquad (6.39)$$

$$\mathbf{R}_P = (x_s \pm x_o, y_s \pm y_o, z_s \pm z_o) \text{ and } \mathbf{R}_N = (lLx, mLy, nLz) \qquad (6.40)$$

$$\delta(t - \tau) = \frac{1}{2\pi} \int_{-\infty}^{\infty} \mathrm{e}^{\mathrm{i}\omega(t-\tau)} d\omega \qquad (1/\mathrm{s}) \qquad (6.41)$$

$$\mathbf{x}_s = (x_s, y_s, z_s), \ \mathbf{x}_o = (x_o, y_o, z_o): \text{ source, observation position} \qquad (6.42)$$

$$L_x, L_y, L_z: \text{ lengths of sides of the room.} \quad (\mathrm{m})$$

The details in the formulation of Eq. (6.36) can be read in the references [1][2][6]. The impulse response is composed of the direct sound followed by the reflections from mirror-image sources.

The frequency response can be derived by substituting a sinusoidal wave function $\mathrm{e}^{\mathrm{i}\omega t}\Delta\omega$, where $\Delta\omega$ indicates $1/\Delta t$ (or taking the Fourier transform of the impulse response) such that [1][2][6]

$$H(\mathbf{x}_s, \mathbf{x}_o, \omega) = Q \sum_{P=1}^{8} \sum_{N=-\infty}^{\infty} \frac{8}{\Lambda_N} \frac{e^{-ik|\mathbf{R}_P + \mathbf{R}_N|}}{4\pi |\mathbf{R}_P + \mathbf{R}_N|}, \quad (\mathrm{m}^2/\mathrm{s}) \tag{6.43}$$

which is nothing but a superposition of the symmetric spherical waves from the point source and its mirror-image sources; however, the frequency response is not intuitively understood from in terms of the room resonances.

On the other hand the frequency response can be formulated by the wave (or modal) theory on the wave equation such that [1][2][8]

$$H(\mathbf{x}_s, \mathbf{x}_o, k) = Q \sum_N \frac{\Lambda_N}{V} \frac{\psi_N(\mathbf{x}_s)\psi_N(\mathbf{x}_o)}{k_N^2 - k^2} \quad (\mathrm{m}^2/\mathrm{s}) \tag{6.44}$$

and

$$p(\mathbf{x}_s, \mathbf{x}_o, k) = i\omega\rho_0 H(\mathbf{x}_s, \mathbf{x}_o, k), \quad (\mathrm{Pa}) \tag{6.45}$$

where ψ_N and k_N denote the N-th modal function and the eigenwave number (1/m), $p(\mathbf{x}_s, \mathbf{x}_o, k)$ (Pa) denotes the sound pressure response, and ρ_0 shows the volume density of the medium (kg/m^3). The N-th modal (or eigen) function is given by

$$\psi_N = \psi_{lmn}(\mathbf{x}) = \cos\frac{l\pi x}{L_x} \cos\frac{m\pi y}{L_y} \cos\frac{n\pi z}{L_z} \tag{6.46}$$

for a rectangular room surrounded by hard walls. The N-th eigenwave number is written as

$$k_N = \sqrt{\left(\frac{l\pi}{L_x}\right)^2 + \left(\frac{m\pi}{L_y}\right)^2 + \left(\frac{n\pi}{L_z}\right)^2}, \quad (1/\mathrm{m}) \tag{6.47}$$

where l, m, n are integers. Substituting $\omega_N = ck_N$,

$$H(\mathbf{x}_s, \mathbf{x}_o, \omega) = c^2 Q \sum_N \frac{\Lambda_N}{V} \frac{\psi_N(\mathbf{x}_s)\psi_N(\mathbf{x}_o)}{\omega_N^2 - \omega^2} \quad (\mathrm{m}^2/\mathrm{s}) \tag{6.48}$$

is obtained, where

$$\omega_N^2 = c^2 k_N^2 = c^2 \left(\left(\frac{l\pi}{L_x}\right)^2 + \left(\frac{m\pi}{L_y}\right)^2 + \left(\frac{n\pi}{L_z}\right)^2\right), \quad (1/\mathrm{s})^2 \tag{6.49}$$

and ω_N(1/s) is called the eigen angular frequency of a rectangular room with rigid walls. Eigenfrequencies, which are interpreted as the frequencies of free vibration, are not harmonic for three-dimensional cases, while they are harmonic for one-dimensional cases. Therefore, structures for musical instruments are one-dimensional, in principle.

6.3.2 Complex frequency, poles and zeros of transfer function

The frequency characteristic function (or frequency response) introduced by Eq. (6.44) can be rewritten as

$$H(\mathbf{x}_s, \mathbf{x}_o, \omega) = -c^2 Q \sum_N \frac{\Lambda_N}{V} \frac{\psi_N(\mathbf{x}_s)\psi_N(\mathbf{x}_o)}{(\omega - \omega_{p1})(\omega - \omega_{p2})} \tag{6.50}$$

$$= -c^2 Q \sum_N \frac{\Lambda_N}{V} \frac{\psi_N(\mathbf{x}_s)\psi_N(\mathbf{x}_o)}{\omega^2 - \omega_N^2 - 2i\omega\delta_N}, \qquad (\mathrm{m}^2/\mathrm{s})$$

where the damping factor is introduced such that

$$\omega_N : N-\text{th eigen angular frequency} \qquad (1/s) \tag{6.51}$$

$$\omega_{N_d} : N-\text{th angular frequency of the free oscillation under the damping condition} \qquad (1/s)$$

$$\omega_{N_d}^2 = \omega_N^2 - \delta_N^2 \quad (1/s)^2 \text{ and } \delta_N : \text{damping factor} \quad (1/s)$$

$$\omega_{p_1} = \omega_{N_d} + i\delta_N \text{ and } \omega_{p_2} = -\omega_{N_d} + i\delta_N. \qquad (1/s). \tag{6.52}$$

Here ω_{p_1} and ω_{p_2} are defined on the complex frequency plane. Taking a pair of parameters $\mathbf{x}_s$ and $\mathbf{x}_o$,

$$H_N(\mathbf{x}_s, \mathbf{x}_o, \omega) = \frac{A_N}{\omega^2 - \omega_N^2 - 2i\omega\delta_N} \tag{6.53}$$

indicates the frequency response of a single-degree-of-freedom system (or a simple resonator), as already developed by Eq. (2.20). The frequency response function defined by Eq. (6.50) can be interpreted as a superposition of the simple resonators' responses.

The angular frequency ω can be extended into the complex frequency plane as

$$\omega^c = \omega + i\delta. \tag{6.54}$$

The frequency response function $H(\mathbf{x}_s, \mathbf{x}_o, \omega^c)$ by extending ω into ω^c is called the transfer function for the impulse response $h(\mathbf{x}_s, \mathbf{x}_o, t)$. The transfer function is defined in the complex frequency plane, while the frequency response function is given on the real frequency axis ω on the complex frequency plane.

The complex (angular) frequency ω_0^c that satisfies

$$H(\mathbf{x}_s, \mathbf{x}_o, \omega_0^c) = 0 \tag{6.55}$$

is called the zero of the transfer function. In contrast, the complex frequency makes

$$H(\mathbf{x}_s, \mathbf{x}_o, \omega_p^c) \to \infty, \tag{6.56}$$

where the transfer function is singular, is called the pole of the transfer function. The complex frequency $\omega_{p_1}, \omega_{p_2}$ defined in Eq. (6.52) are the examples of the poles.

Poles are defined independent of the locations of the source and observation points, while zeros depend on locations of the source and observation points [9][10]. The transfer function is not well defined at the poles; however, the poles are located only above the real frequency axis on the complex frequency plane for stable systems, in which the impulse response is stable [1]. The poles below the real frequency axis make the system unstable, for example during the howling of a feedback system [11]. A single resonator has a pair of poles $\omega_{p_1}, \omega_{p_2}$; however, there is no zero for the transfer function of the single resonator. The zeros are typical complex frequencies for multi-degree-of-freedom systems [9][10].

6.3.3 Number of eigenfrequencies and modal density

The poles of transfer functions are closely related to the eigenfrequencies. The eigenfrequencies of a rectangular room are located in the wave number space, as shown in Fig. 6.3.

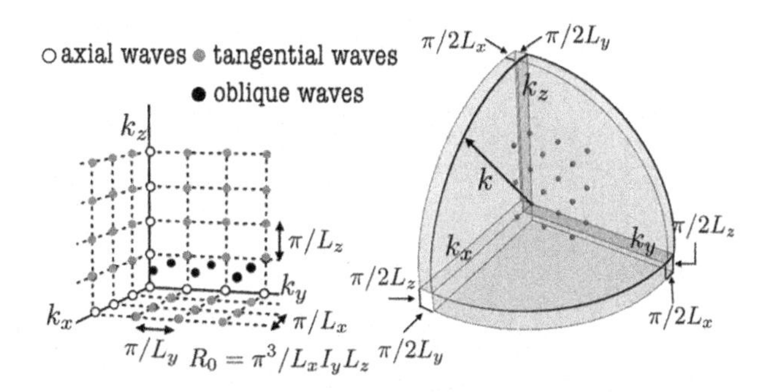

FIGURE 6.3

Left panel: Eigenfrequencies indicated by lattices in wave number space for a rectangular room. Right: Additional volumes for counting numbers of tangential and axial eigenfrequencies in wave number space.

The number of eigenfrequencies lower than ω is representative of the transfer function for a multi-degree-of-freedom system [1][2][8]. The eigenfrequencies are located at lattices in the wave number space, as shown in Fig. 6.3. The volume of a single lattice in the wave number space is

$$R_0 = \frac{\pi}{L_x} \cdot \frac{\pi}{L_y} \cdot \frac{\pi}{L_z} = \frac{\pi^3}{V}, \quad (1/\text{m}^3) \tag{6.57}$$

where $V = L_x L_y L_z$ (m^3). Taking the volume of a 1/8 sphere with radius k in the wave number space, the ratio $N_3(k) = R_3(k)/R_0$ gives an estimate of the number of eigenfrequencies lower than k, where $R_3(k)$ denotes the volume of a 1/8 sphere with radius k in the wave number space. The eigenfrequencies located at the lattices where one of the integers (l, m, n) is 0 are called the tangential eigenfrequencies, or eigenfrequencies of tangential modes. Volumes to be occupied by the lattices indicating the tangential eigenfrequencies, however, are not contained fully in the 1/8 sphere. To count all of the tangential eigenfrequencies, the volumes are added, as shown in the right panel of Fig. 6.3, to the 1/8 sphere [1][2][8]. The number to be added for the tangential waves, $N_2(k)$, is estimated by

$$N_2(k) \cong \frac{R_2(k)}{R_0} = \frac{\left(\frac{\pi}{2L_x} + \frac{\pi}{2L_y} + \frac{\pi}{2L_z}\right) \cdot \frac{\pi k^2}{4}}{R_0} = \frac{Sk^2}{16\pi}, \tag{6.58}$$

where

$$S = 2(L_x L_y + L_y L_z + L_z L_x) \qquad (\mathrm{m}^2) \tag{6.59}$$

denotes the surface area of the rectangular room.

The eigenfrequencies for the lattices where two of the integers l, m, n are zeros are called axial eigenfrequencies or eigenfrequencies of axial modes. According to the right panel of Fig. 6.3 once more, the number to be added $N_1(k)$ is estimated such that

$$N_1(k) \cong \frac{R_1(k)}{R_0} = \frac{\left(\frac{\pi}{2L_x}\frac{\pi}{2L_y} + \frac{\pi}{2L_y}\frac{\pi}{2L_z} + \frac{\pi}{2L_z}\frac{\pi}{2L_x}\right) \cdot k}{R_0} = \frac{Lk}{16\pi}, \tag{6.60}$$

where

$$L = 4(L_x + L_y + L_z). \qquad (\mathrm{m}) \tag{6.61}$$

An estimate of the total number of eigenfrequencies can be written as

$$N(k) \cong N_3(k) + N_2(k) + N_1(k) = \frac{V}{6\pi^2}k^3 + \frac{S}{16\pi}k^2 + \frac{L}{16\pi}k. \tag{6.62}$$

The result of Eq. (6.62) also gives an estimation for the number of eigenfrequencies in a nonrectangular room [1][2][8]. Taking only the eigenfrequencies of oblique modes where none of the integers (l, m, n) is zero, the number of oblique eigenfrequencies can be estimated by subtracting the number of tangential eigenfrequencies such that

$$N_{ob}(k) \cong N(k) - N_{tan}(k) \cong \frac{V}{6\pi^2}k^3 - \frac{S}{16\pi}k^2 + \frac{L}{16\pi}k, \tag{6.63}$$

where

$$N_{tan}(k) \cong 2\frac{R_2(k)}{R_0} = \frac{S}{8\pi}k^2. \tag{6.64}$$

Fig. 6.4 shows an example of counting oblique eigenmodes less than the wave number k for a rectangular room shown by the right panel in the figure.

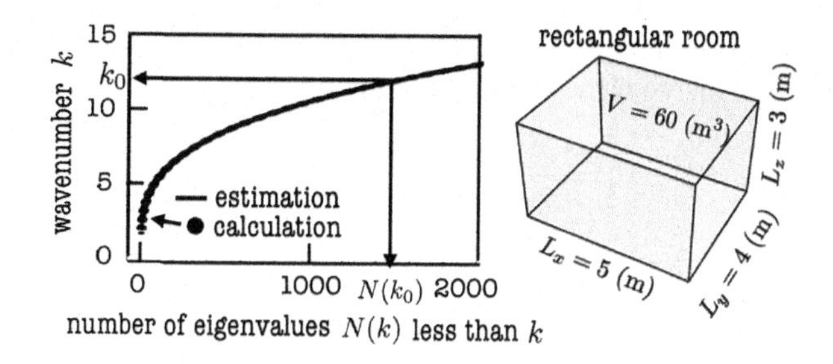

FIGURE 6.4

Number of oblique eigenmodes less than wave number k in rectangular room shown in the right panel.

The numerical result indicated by the solid circles mostly follows the estimation in Eq. (6.63) given by the solid line.

Similarly to Eq. (6.63) the number of xy-tangential eigenfrequencies where neither of the integers (l, m) is zero is estimated by

$$N_{tan_{xy}} \cong \frac{\frac{\pi k^2}{4}}{\frac{\pi^2}{S_{xy}}} - \frac{\left(\frac{\pi}{2L_x} + \frac{\pi}{2L_y}\right)k}{\frac{\pi^2}{S_{xy}}} = \frac{S_{xy}k^2}{4\pi} - \frac{L_x + L_y}{2\pi}k, \tag{6.65}$$

where $S_{xy} = L_x L_y$ (m^2). The number of x-axial eigenfrequencies is given by

$$N_{ax_x} \cong \frac{k}{\frac{\pi}{L_x}} = \frac{L_x}{\pi}k. \tag{6.66}$$

The distribution of eigenmodes in a nonrectangular two-dimensional space follows Eq. (6.65) [1][2]. Substituting $k = \omega/c$, the modal density can be estimated as

$$n_p(\omega) = \frac{dN(\omega)}{d\omega} \cong \frac{dN_3(\omega)}{d\omega} = \frac{V\omega^2}{2\pi^2 c^3}, \qquad \text{(s)} \tag{6.67}$$

where $N_3 = R_3/R_0$ and R_3 denotes the volume of a 1/8 sphere with radius k. The modal density increases in proportion to ω^2 in three-dimensional space.

6.3.4 Modal bandwidth and modal overlap

A response at the steady state of a multi-degree-of-freedom system is a superposition of the responses by resonators as shown by Eq. (6.50). Suppose a response of a single resonator

$$Y(\omega) = \frac{-1}{\omega^2 - \omega_0^2 - 2i\omega\delta} \tag{6.68}$$

following Eq. (2.20), where ω_0 denotes the eigen angular frequency of the resonator, and δ is the damping factor. The response of a single resonator is called the modal response, which makes the response of a multi-degree-of-freedom system by superposition. Take the squared response $|Y|^2$ such that

$$|Y(\omega)|^2 = \frac{1}{(\omega^2 - \omega_0^2)^2 + 4\omega^2\delta^2}, \quad (\mathrm{s}^4) \tag{6.69}$$

which shows a magnitude of a squared resonance response. The modal bandwidth B_M of the resonant response is defined as [2][12][13]

$$\int_0^\infty |Y(\omega)|^2 d\omega \cong \frac{\pi\delta}{4\omega_0^2\delta^2} = \frac{1}{4\omega_0^2\delta^2} \cdot B_M, \quad (\mathrm{s}^3) \tag{6.70}$$

as shown in Fig. 6.5, where $B_M = \pi\delta$ (1/s).

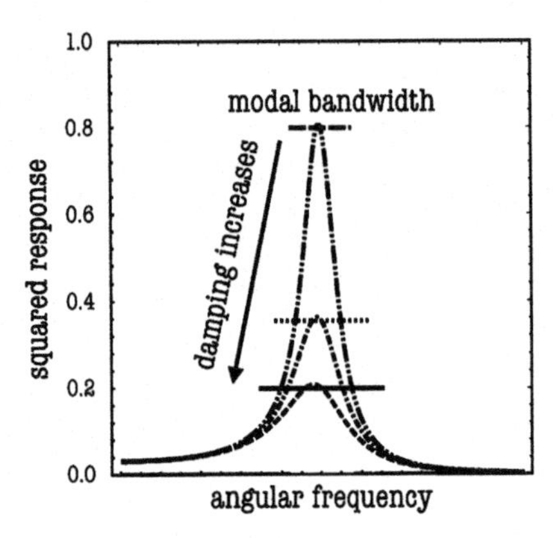

FIGURE 6.5

Squared magnitudes of modal responses and modal bandwidths.

The modal bandwidth is narrower (wider) as the damping factor decreases (increases).

The modal overlap can be an indicator of how densely the modal response is superposed by

$$M(\omega) = n_p(\omega)B_M, \tag{6.71}$$

where $n_p(\omega)$ denotes the modal density. The modal overlap $M(\omega)$ shows an estimate of how many modes are simultaneously excited in a sound field when a single sinusoidal wave is radiated from a source into the field.

Eq. (6.50) can be rewritten as an approximating formula for the squared response by averaging the sound source and receiving positions through the space [2][7][8][14]. Taking the ensemble average with respect to source and receiving positions, then Eq. (6.50) becomes

$$\overline{|H(\omega)|^2} = \frac{c^4 Q^2}{V^2} \sum_N \frac{1}{(\omega^2 - \omega_N^2)^2 + 4\omega^2\delta_N^2} \quad (\mathrm{m}^2/\mathrm{s})^2 \tag{6.72}$$

because of orthogonality of the modal functions. The summation in Eq. (6.72) can be decomposed of the oblique, tangential, and axial waves such that

$$|\overline{H}(\omega)|^2 = |\overline{H}_{ob}(\omega)|^2 + |\overline{H}_{tan}(\omega)|^2 + |\overline{H}_{ax}(\omega)|^2. \qquad (\mathrm{m}^2/\mathrm{s})^2 \qquad (6.73)$$

Introducing the integral formula in Eq. (6.70), and taking the number of superposed modes estimated by $M/2$ where the phase difference is expected within $\pi/2$ [7][8], then

$$|\overline{H}_{ob}(\omega)|^2 \cong \frac{c^4 Q^2}{V^2}\frac{M_{ob}}{2}\frac{1}{4\omega^2\delta_{N_{ob}}^2} \qquad (\mathrm{m}^2/\mathrm{s})^2 \qquad (6.74)$$

$$|\overline{H}_{tan}(\omega)|^2 \cong \frac{c^4 Q^2}{V^2}\sum_{xy}\frac{M_{xy}}{2}\frac{1}{4\omega^2\delta_{N_{xy}}^2}$$

$$|\overline{H}_{ax}(\omega)|^2 \cong \frac{c^4 Q^2}{V^2}\sum_{x}\frac{M_{x}}{2}\frac{1}{4\omega^2\delta_{N_{x}}^2} \qquad (6.75)$$

are obtained. According to the modal density and modal overlap,

$$n_{ob}(\omega) \cong \frac{V\omega^2}{2\pi^2 c^3}\left(1 - \frac{\pi S c}{4V\omega} + \frac{\pi L c^2}{8V\omega^2}\right) \qquad (\mathrm{s}) \qquad (6.76)$$

$$n_{xy}(\omega) \cong \frac{L_x L_y \omega}{2\pi c^2}\left(1 - \frac{(L_x + L_y)c}{L_x L_y \omega}\right) \quad \text{and} \quad n_x = \frac{L_x}{\pi c}, \qquad (6.77)$$

where

$$M_{ob} = \pi\delta_{N_{ob}} n_{ob}, \quad M_{xy} = \pi\delta_{N_{xy}} n_{xy}, \quad \text{and} \quad M_x = \pi\delta_{N_x} n_x, \qquad (6.78)$$

the space-averaged squared responses in Eq. (6.74) are rewritten as

$$|\overline{H}_{ob}(\omega)|^2 \cong \frac{cQ^2}{16\pi V}\frac{1}{\delta_{N_{ob}}}\left(1 - \frac{\pi S c}{4V\omega} + \frac{\pi L c^2}{8V\omega^2}\right) \qquad (\mathrm{m}^2/\mathrm{s})^2 \qquad (6.79)$$

$$|\overline{H}_{tan}(\omega)|^2 \cong \frac{cQ^2}{16\pi V}\sum_{xy}\frac{1}{\delta_{N_{xy}}}\frac{\pi c L_x L_y}{V\omega}\left(1 - \frac{(L_x + L_y)c}{L_x L_y \omega}\right) \qquad (6.80)$$

$$|\overline{H}_{ax}(\omega)|^2 \cong \frac{cQ^2}{16\pi V}\sum_{x}\frac{1}{\delta_{N_x}}\frac{2\pi c^2 L_x}{V\omega^2}. \qquad (6.81)$$

Recalling the sound power output P_0 (W) of a point source in a free field such that [2][15]

$$Q^2 = \frac{8\pi c}{\rho\omega^2}P_0, \qquad (\mathrm{m}^3/\mathrm{s})^2 \qquad (6.82)$$

then

$$\omega^2\rho^2\frac{cQ^2}{16\pi} = \frac{P_0\rho c^2}{2} \qquad (\mathrm{Pa}^2 \cdot \mathrm{m}^3/\mathrm{s}) \qquad (6.83)$$

is derived. Consequently, the space-averaged squared response can be given by

$$\frac{|\overline{p}(\omega)|^2}{4\rho c P_0} \cong \frac{1}{A_{ob}}\left(1 - \frac{\pi Sc}{4V\omega} + \frac{\pi Lc^2}{8V\omega^2}\right) + \sum_{xy}\frac{\pi c L_x L_y}{\frac{4}{\pi}A_{xy}V\omega}\left(1 - \frac{c(L_x + L_y)}{L_x L_y \omega}\right) + \sum_{x}\frac{2\pi c^2 L_x}{2A_x V\omega^2}, \quad (1/\text{m}^2) \tag{6.84}$$

where

$$\omega^2\rho^2\left|\overline{H}(\omega)\right|^2 = |\overline{p}(\omega)|^2 \quad (\text{Pa}^2) \tag{6.85}$$

and

$$2\delta_{N_{ob}} = \frac{cA_{ob}}{4V}, \quad 2\delta_{N_{xy}} = \frac{cA_{xy}}{\pi V}, \quad \text{and} \quad 2\delta_{N_x} = \frac{cA_x}{2V} \quad (1/\text{s}) \tag{6.86}$$

according to the relationship between the decaying constant δ and equivalent absorption area A [2][7][8][16]. The squared sound pressure response after averaging the source and receiving positions is represented by superposition of the oblique, tangential, and axial wave modal responses. The equivalent absorption area is originally defined in terms of the geometrical acoustics represented by the reverberation time. The superposed representation in Eq. (6.84) can be understood as a hybrid formulation of the wave theory and geometric acoustics that are applicable to nonrectangular rooms using geometrical parameters through appropriate selections of oblique, tangential, or axial waves. The expression given in Eq. (6.84) is also derived from the geometrical acoustics [2][7].

6.3.5 Reverberation formula as superposition of oblique, tangential, and axial wave fields

The frequency response at the steady state for the space-averaged squared sound pressure is given by Eq. (6.84), where the response is a superposition of the oblique, tangential, and axial waves. The superposed formula indicates that the reverberation energy decay does not follow an exponential decay curve after the sound source stops because the decay constants depend on the modal types, as shown in Eq. (6.86). The reverberation decay formula can be written

$$\frac{|\overline{p}(\omega,t)|^2}{4\rho c P_0} \cong \frac{1}{A_{ob}}\left(1 - \frac{\pi Sc}{4V\omega} + \frac{\pi Lc^2}{8V\omega^2}\right)e^{-\frac{cA_{ob}}{4V}t} + \sum_{xy}\frac{\pi c L_x L_y}{\frac{4}{\pi}A_{xy}V\omega}\left(1 - \frac{c(L_x + L_y)}{L_x L_y \omega}\right)e^{-\frac{cA_{xy}}{\pi V}t} + \sum_{x}\frac{2\pi c^2 L_x}{2A_x V\omega^2}e^{-\frac{cA_x}{2V}t}, \quad (1/\text{m}^2) \tag{6.87}$$

following Eq. (6.84) [2][7][8][16][17].

The reverberation formula is a superposition of the three types of decay curves for the oblique, tangential, and axial wave fields. The reverberation time can be defined by [2][7][16][17][18]

$$TR_{ob} = TR_3 \cong 0.163\frac{V}{A_3} \quad \text{(s)} \tag{6.88}$$

$$TR_{tan} = TR_2 \cong 0.128\frac{S_2}{A_2} \tag{6.89}$$

$$TR_{ax} = TR_1 \cong 0.041\frac{L_1}{A_1}, \tag{6.90}$$

where

$$V = L_x L_y L_z \quad (\text{m}^3), \quad A_3 = -\log_e(1-\alpha)S \quad (\text{m}^2) \tag{6.91}$$

$$S_2 = L_x L_y \quad (\text{m}^2), \quad A_2 = -\log_e(1-\alpha)L_2 \quad (\text{m}) \quad L_2 = 2(L_x + L_y) \quad (\text{m}) \tag{6.92}$$

$$L_1 = L_x \quad (\text{m}), \quad A_1 = -\log_e(1-\alpha). \tag{6.93}$$

The reverberation energy decays following a nonexponential decay curve because of superposition of exponential functions with different damping constants in general [16][17][19]. The initial decay (or the shorter reverberation time for the initial portion) is due to the oblique waves that have rapid decay constants because of sound absorption by total areas of surrounding walls. The dominant portion of the decay curve expresses the longer reverberation time of the tangential waves that travel in the plane parallel to the floor and ceiling surrounded by the hard side walls, for example. High intelligibility of speech is expected due to the rapid initial decay, even under long reverberation conditions, as will be described in the next chapter. Another example of the frequency characteristics of reverberation time for the mixed fields are presented in detail in the references [2][7][17][20].

6.4 Minimum-phase and all-pass components for transfer functions

An impulse response can be defined for a sound field in a room, as described by Eq. (6.36). The impulse response can be interpreted in terms of discrete linear systems. An artificial reverberator is the classic example of sound simulators using all-pass systems [21]. The notion of minimum-phase and all-pass systems are the keys to characterizing the sound field conditions, such as the distance between the source and receiving positions, as will be described in Chapter 8. This section introduces the transfer functions for discrete systems in terms of the poles and zeros or the magnitude and phase responses.

6.4.1 Discrete systems and Fourier transforms

The transfer function can be defined for discrete systems as well. Suppose an impulse response sequence $h(n)$. The $z-$transform of the sequence $h(n)$ is defined as

$$H(z^{-1}) = \sum_n h(n)z^{-n} \quad \text{and} \quad z = re^{i\Omega}, \tag{6.94}$$

where z is a complex variable. Setting $r = 1$,

$$H(e^{-i\Omega}) = \sum_n h(n)e^{-in\Omega} \tag{6.95}$$

is called the Fourier transform of the sequence $h(n)$, while $H(z^{-1})$ is called the transfer function of the impulse response $h(n)$. The Fourier transform $H(e^{-i\Omega})$ is called the frequency characteristics of the impulse response [1][11].

Reverberation is significant to sound quality in a listening space such as a concert hall; however, the initial echoes or reflected sounds that come to a listener following the direct sound before the reverberation, are quite impressive to the listener [22]. Suppose an impulse response including a single echo:

$$h(n) = \begin{cases} 1 & n = 0 \\ a & n = N \geq 1\,. \\ 0 & others. \end{cases} \tag{6.96}$$

The $z-$ and Fourier-transforms of $h(n)$ become

$$H(z^{-1}) = 1 + az^{-N} \quad \text{and} \quad H(e^{-i\Omega}) = 1 + ae^{-iN\Omega}, \tag{6.97}$$

respectively, where Ω denotes the normalized angular frequency, N indicates the time delay of the reflection sound following the direct sound by the time delay of $\tau = N \cdot T_s$ (s), and T_s (s) denotes the period of sampling.

Introducing the sampling period T_s, the normalized angular frequency Ω is

$$\Omega = \omega T_s = \frac{2\pi F}{F_s}, \qquad \text{(rad)} \tag{6.98}$$

where $F_s = 1/T_s$ denotes the sampling frequency (Hz), and Ω approaches 2π when $\omega = 2\pi F_s$. Eq. (6.97) shows a complex function of a complex variable $e^{-iN\Omega}$ and gives the output response of the linear system defined by Eq. (6.96) to a sinusoidal signal. Fig. 6.6 shows an illustration of Eq. (6.97) for $N = 1$, where the left, center, and right panels show the cases, $a = 0.5, 1, 2$, respectively. The dotted line in the center panel shows that the unit circle moves by $+1$ to the right on the real axis after the Fourier transform. Both the original $e^{-i\Omega}$ and the response $H(e^{-i\Omega})$ are periodic with respect to Ω by the period 2π. The magnitude of the response $|H(e^{-i\Omega})|$ becomes 0 for $\Omega = \pi$, and takes the maximum ($= 2$) for $\Omega = 0$.

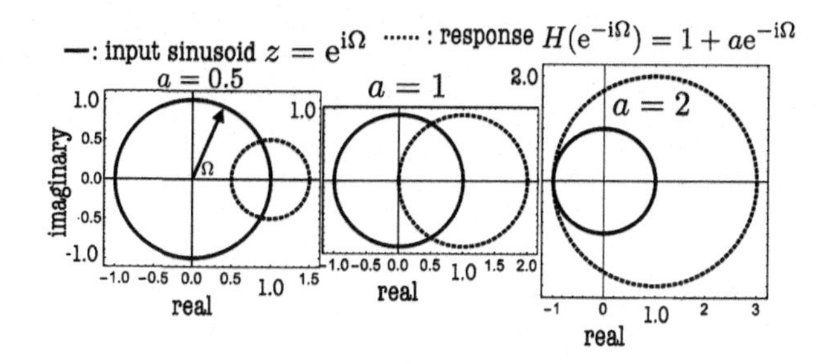

FIGURE 6.6

Fourier transforms from Eq. (6.97) for $a = 0.5, 1, 2$ and $N = 1$.

The periods regarding Ω are 2π; however, the phase angle or argument of the responses are different from each other. The argument is accumulated from 0 to 2π for the original variable, when Ω moves from 0 to 2π. The real part of the response (dotted circle) for $a = 1$ (center panel) or $a = 0.5$ (left panel) is nonnegative, even when Ω moves from 0 to 2π. As shown in the panels, the phase angle returns to 0 (initial position), even when Ω moves from 0 to 2π. In contrast, the angle for the response when $a = 2$ (right panel) is accumulated to 2π. The original angle is also accumulated to 2π (unit circle) before the Fourier transform.

There are remarkable differences in the three panels, although all of the panels are periodic with the period 2π. In addition, no zeros can be seen in the left and right panels because there is no Ω_0 that satisfies

$$1 + ae^{-i\Omega_0} = 0 \tag{6.99}$$

when $a \neq \pm 1$.

Fig. 6.7 shows the case when $N = 2$ for Eq. (6.97). The period of the response is $2\pi/2 = \pi$, which indicates $2\pi/N$ when $N = 2$.

The magnitude (or absolute) responses $|H(e^{-i\Omega})|$ illustrated by the bottom panel confirm the zeros, minima and maxima, and the period. A single echo, whose Fourier transform is expressed by Eq. (6.97), gives severe spectral deformation on the frequency responses. The spectral deformation is periodic and its period depends on the delay time of the single echo following the direct sound. The same type of spectral deformation can occur when a recording microphone is located close to a rigid or reflective wall.

6.4.2 Zeros and z–transforms of discrete systems

Fig. 6.6 showed no zeros when $a \neq 1$. The z–transform could find the zero instead of the Fourier transform. Recall that zeros are not found on the real variable x for the equation

$$x^2 + 2x + 3 = 0. \tag{6.100}$$

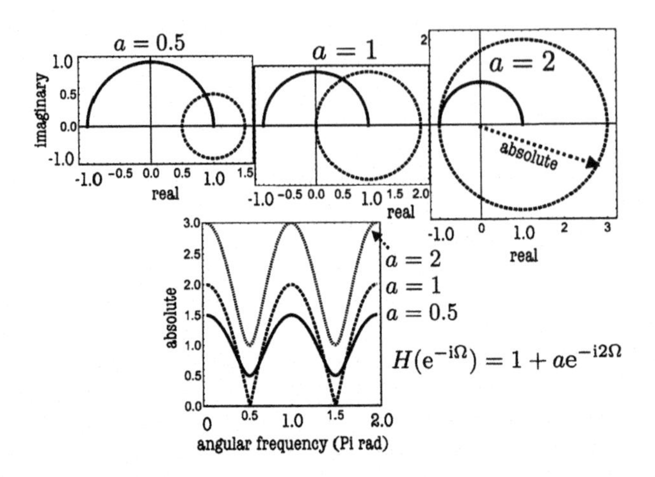

FIGURE 6.7

Similar to Fig. 6.6, but for $N = 2$, and the bottom panel shows the absolute of Eq. (6.97).

The zeros for the z−transform that satisfy

$$H(z_0^{-1}) = 1 + az_0^{-N} = 0 \tag{6.101}$$

can be expressed as

$$z_0^{-N} = -a^{-1} = a^{-1}e^{i\pi}, \tag{6.102}$$

where a is a positive real number. Setting

$$z_0 = |z_0|e^{i\Omega_0} = r_0 e^{i\Omega_0}, \tag{6.103}$$

then

$$r_0^{-N} e^{-iN\Omega_0} = a^{-1}e^{i\pi}. \tag{6.104}$$

Consequently,

$$r_0 = a^{1/N} \quad \text{and} \quad \Omega_0 = -\pi/N + 2m\pi, \tag{6.105}$$

which confirm

$$1 + az_0^{-N} = 1 + a \cdot (a^{1/N}e^{-i\pi/N})^{-N} = 0, \tag{6.106}$$

where m is an integer. The argument of the zero Ω_0 is independent of $|a|$, but depends on the argument of a (positive or negative in the example). In contrast, the magnitude of the zero is

$$|z_0| = a^{1/N} \begin{cases} > 1 & a > 1 \\ \leq 1 & a \leq 1 \end{cases}, \tag{6.107}$$

which depends on $|a|$ and N. The zeros are located on the complex plane; they are located outside the unit circle when $a > 1$, while they are found inside for $a < 1$ or on the unit circle when $a = 1$.

The zeros in Eq. (6.106) explain why no zero can be seen when $a = 0.5$ or $a = 2$ in Fig. 6.7. The Fourier transform is defined on the unit circle in the complex plane, while the domain of the definition of the z−transform is the complex plane, except for the singularities. The zeros on the unit circle can be seen by the magnitude of the Fourier transform; however, the other zeros cannot be found for the Fourier transform because they are not on the unit circle.

Magnitude (absolute) response makes the troughs at the zeros. Fig. 6.8 shows the magnitude response for

$$H(z^{-1}) = (1 + 0.5z^{-1})(1 + z^{-1})(1 + 2z^{-1}), \tag{6.108}$$

where the variable $|z| = r$ is taken in the interval $0.4 \le r \le 3$, keeping $\Omega = \Omega_0 = -\pi$.

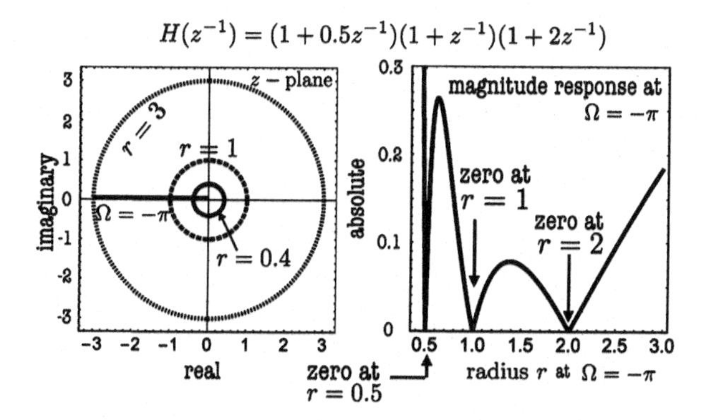

FIGURE 6.8

Troughs of magnitude for Eq. (6.108) at $\Omega = -\pi$ for $a = 0.5, 1, 2$ and $N = 1$.

The troughs become shallower as $|z_0| = a$ is farther from $|z|$ at which the z−transform is taken. In contrast, they are deeper when r_0 is closer to $|z|$. The frequency of the trough gives the frequency of the zero, and the depth gives an estimate of the radius of the zero location $|z_0| = r_0$.

6.4.3 Zeros and minimum-phase transfer function

Return to Eq. (6.97) and Fig. 6.6 once more. The phase is accumulated to -2π for $a = 2$, while no accumulation is seen for $a = 1$ or $a = 0.5$. This is understood according to the mathematical formula [23]

$$\int_c \frac{f'}{f} = \left[\log[|f(\Omega)|e^{i\phi(\Omega)}]\right]_0^{2\pi} = 2\pi i(N_z^- - N_p), \tag{6.109}$$

where f is the z–transform, f' is the first derivative of f, c shows the circle at the center $z = 0$ with the radius r, N_z^- denotes the number of the zeros inside the circle c, and N_p gives the number of the poles (or singularities) inside the circle. The singularity or the pole for Eq. (6.97) is located at the origin of the complex plane. When $a = 0.5$, the zero is inside the unit circle ($r = 1$) on which the Fourier transform is taken, and thus

$$\Phi(2\pi) = \phi(2\pi) - \phi(0) = 2\pi(N_z^- - N_p) = 0. \tag{6.110}$$

On the other hand, when $a = 2$ the zero is outside the unit circle, and thus the accumulated phase becomes -2π.

Setting $r = 0.25$ as the radius of the circle on which the z–transforms are taken, then the accumulated phase responses yield -2π for all of the cases for $a = 0.5, 1, 2$ as illustrated in the left panel of Fig. 6.9 because all the zeros are located outside the circle. In contrast, the accumulated phases are 0 for $r = 2.5$, as shown in the right panel of Fig. 6.9, since no zeros are outside the circle.

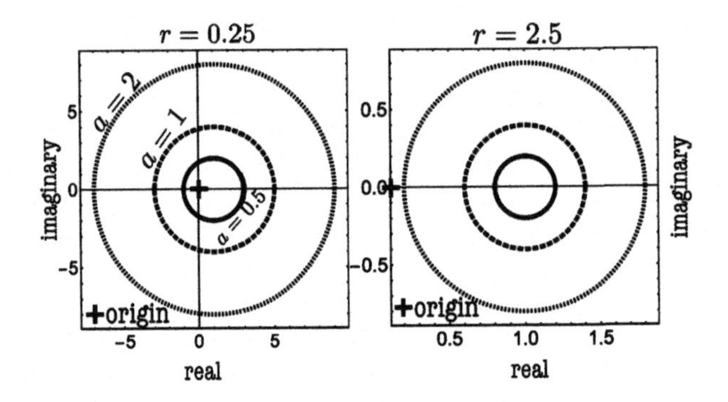

FIGURE 6.9

z–transforms at $r = 0.25$ and 2.5 for $a = 0.5, 1, 2$ and $N = 1$.

The accumulated phase estimated by Eq. (6.109) shows how many times the response goes around the origin in the z–plane, when the angular frequency moves from 0 to 2π.

The z–transforms with the variable r give estimates of the zero locations, whether inside or outside the circle of radius r, by counting the accumulated phase [24]. The Fourier transform makes no phase accumulation when no zeros are outside the unit circle. A sequence whose z–transform has no zeros outside the unit circle is called the minimum-phase sequence.

Return to the definition of z–transform in Eq. (6.94). Applying an exponential windowing function r^{-n} to the time sequence $h(n)$ such that

$$\hat{h}(n) = w(n)h(n) = r^{-n}h(n), \tag{6.111}$$

the Fourier transform of $\hat{h}(n)$ becomes

$$\hat{H}(e^{-i\Omega}) = \sum_n r^{-n} h(n) e^{-i\Omega n} = \sum_n h(n) z^{-n} \text{ for } z = re^{i\Omega}. \tag{6.112}$$

The $z-$transform gives the Fourier transform of the sequence defined by applying the exponential windowing $w(n) = r^{-n}$. The exponential windowing increases the decaying speed of the time sequence $h(n)$ for $r > 1$.

The sound timber of a listening room often changes when sound absorbing materials are set into the room. The zeros making the troughs for the magnitude spectral response would explain some of the reasons why the timber could change. Some zeros might be located far away from the unit circle, and thus the troughs due to the zeros may not be prominent under ordinary room conditions. However, the troughs might be prominent when the decaying condition (or reverberation time) of the room changes by introducing sound absorbers into the room. Under the high-damping conditions, the far-away zeros can be close to the unit circle on which the Fourier transform is taken. As the zeros come closer to the unit circle, the zeros are prominent and the troughs become deeper.

Fig. 6.10 is an example of the troughs due to the zeros changing by the exponential windowing for a small room [2][24][25][26].

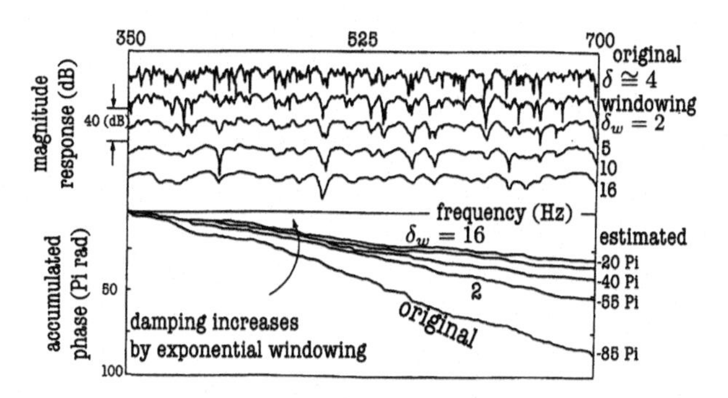

FIGURE 6.10

Magnitude and phase frequency responses after exponential windowing from Fig. 6 [25].

Some prominent zeros can be seen when the decaying constant becomes larger by the exponential windowing, and the accumulated phase decreases as the decaying constant increases because the number of zeros inside the unit circle increases. The trend of accumulated phase can be estimated according to Eq. (6.109). The details for the phase estimation can be seen in the references [2][24][25].

The example shown by Fig. 6.10 implies that the minimum-phase system that is made by increasing the damping yields a shorter record length of the impulse response. This shorter record length of the response is necessary when designing loudspeaker systems where a magnitude frequency response is given. It might explain why a minimum-phase response would be a desirable goal of loudspeaker design.

6.4.4 Poles and all-pass systems

Suppose a transfer function

$$H(z^{-1}) = 1 + az^{-N}, \quad (6.113)$$

where $|a| < 1$. The function given by Eq. (6.113) has zeros

$$z_0 = a^{1/N} e^{-i\pi/N}, \quad (6.114)$$

as described by Eq. (6.105). Setting $N = 1$ for simplicity, the function

$$G(z^{-1}) = \frac{1}{H(z^{-1})} = \frac{1}{1 + az^{-1}} \quad (6.115)$$

is called the inverse system for $H(z^{-1})$, where $|a| < 1$. The zero for $H(z^{-1})$ makes the pole of $G(z^{-1})$ such that

$$G(z_0^{-1}) = \frac{1}{H(z_0^{-1})} \to \infty, \quad (6.116)$$

where

$$H(z_0^{-1}) = 0. \quad (6.117)$$

In other words the zero for $H(z^{-1})$, z_0, is the pole z_p for $G(z^{-1})$.

The transfer function $G(z^{-1})$ that defines the inverse system of $H(z^{-1})$ can be rewritten as

$$G(z^{-1}) = \frac{1}{1 + az^{-1}} = \lim_{N \to \infty} \sum_{n=0}^{N} a^n z^{-n}, \quad (6.118)$$

where

$$|a| < 1, \quad |z| > |a| \quad (6.119)$$

for the convergence of the series in Eq. (6.118). The unit circle is included in the region of the convergence under the condition in Eq. (6.119). Taking $z = e^{i\Omega}$ on the unit circle,

$$G(e^{-i\Omega}) = \frac{1}{1 + ae^{-i\Omega}} = \lim_{N \to \infty} \sum_{n=0}^{N} a^n e^{-i\Omega n} \quad (6.120)$$

holds because

$$|a||e^{-i\Omega}| < 1 \quad (6.121)$$

meets the convergence condition. Thus, the impulse response is well defined as $g(n) = a^n$ under the convergence condition. The convergence condition requires that the pole of the inverse system must be located inside the unit circle for the stable impulse response. A minimum-phase system has a stable inverse system, because the transfer function of a minimum-phase system has no zeros outside the unit circle.

The left panel of Fig. 6.11 is an illustration of $G(e^{-i\Omega})$ in Eq. (6.120).

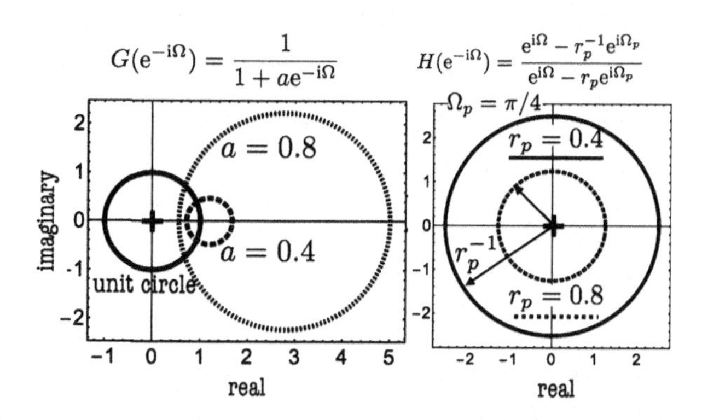

FIGURE 6.11

Fourier transforms from Eqs. (6.120) (left panel) and (6.129) (right panel) where $a = 0.4, 0.8$ (left), $r_p = 0.4, 0.8$, and $\Omega_p = \pi/4$ (right).

The Fourier transform $G(e^{-i\Omega})$ keeps the circle except the radius and the center. The enlarged radius is due to the pole effect on the magnitude. The unit circle is plotted for the reference. The circle for $G(e^{-i\Omega})$ is centered at

$$\frac{1}{2}\left(\frac{1}{1-|a|} + \frac{1}{1+|a|}\right) = \frac{1}{1-|a|^2}. \tag{6.122}$$

Define $F(z^{-1})$ by subtraction such that

$$F(z^{-1}) = G(z^{-1}) - \frac{1}{1-|a|^2} = \frac{-|a|^2}{1-|a|^2} \cdot \frac{1+\frac{1}{a^*}z^{-1}}{1+az^{-1}}, \tag{6.123}$$

where $G(z^{-1})$ is given by Eq. (6.118). Substituting $z = e^{i\Omega}$ and taking the magnitude, then

$$|F(e^{-i\Omega})| = \left|\frac{-|a|^2}{1-|a|^2}\right| \cdot \left|\frac{e^{-i\Omega}(e^{i\Omega} + a^{*-1})}{a(e^{-i\Omega} + a^{-1})}\right| = \frac{|a|}{1-|a|^2}, \tag{6.124}$$

which explains why $G(e^{-i\Omega})$ makes a circle with radius $|a|/(1-|a|^2)$ centered at $1/(1-|a|^2)$ when $|a| < 1$.

The left panel of Fig. 6.11 shows $G(e^{-i\Omega})$ produces no accumulated phase. This is true because the transfer function $G(z^{-1})$ given by Eq. (6.118) has a zero at the origin $z = 0$, and also a pole at $z = -a$. The pair of the pole and the zero cancels the phase accumulation according to Eq. (6.109) so that no phase can be accumulated.

Suppose a transfer function

$$H(z^{-1}) = \frac{1 - \beta z^{-1}}{1 - \alpha z^{-1}}, \tag{6.125}$$

where α and β are complex numbers. The pole and zero are given by

$$z_0 = \beta, \quad z_p = \alpha. \tag{6.126}$$

An interesting case would be when $\beta = \alpha^{-1^*} = \alpha^{*^{-1}}$. Setting $z_p = \alpha = r_p e^{i\Omega_p}$ for $r_p < 1$, then the zero becomes

$$z_0 = \beta = r_p^{-1}(e^{-i\Omega_p})^* = r_p^{-1} e^{i\Omega_p}. \tag{6.127}$$

The relationship between z and $z^{*^{-1}}$ for a complex number z is called the inversion of a complex number (see Fig. 1.2). The transfer function can be rewritten as

$$H(z^{-1}) = \frac{1 - \beta z^{-1}}{1 - \alpha z^{-1}} = \frac{z - z_0}{z - z_p} = \frac{z - z_p^{-1^*}}{z - z_p} \tag{6.128}$$

by using the pair of pole and zero. Take the Fourier transform

$$H(e^{-i\Omega}) = \frac{e^{i\Omega} - r_p^{-1} e^{i\Omega_p}}{e^{i\Omega} - r_p e^{i\Omega_p}}. \tag{6.129}$$

The right panel of Fig. 6.11 illustrates Eq. (6.129). The function preserves the circle centered at the origin of the complex plane, except for the radius. The reservation of the circle is confirmed by taking the magnitude such that

$$|H(e^{-i\Omega})| = r_p^{-1}. \tag{6.130}$$

The magnitude indicates the circle of the radius $1/r_p$ centered at the origin so that the magnitude response of the Fourier transform might be constant independent of the angular frequency Ω, even when Ω changes from 0 to 2π.

A system composed of inversion pairs of the poles and zeros is called the all-pass system when the poles are located inside the unit circle. The magnitude response is constant independent of the frequency; however, the phase is frequency dependent and the accumulated phase is produced according to the difference of the numbers of the poles and zeros inside the unit circle. The accumulated phase is -2π in the example in Eq. (6.128) because a single pole is located at z_p inside the unit circle, while the zero, z_0, is outside the unit circle, and thus the pair of pole and zero does not cancel the phase accumulation to each other.

All-pass systems are historically used for a reverberator in an application of audio-engineering [21]. A pole makes the impulse response longer (or reverberation longer) as indicated by Eq. (6.118); however, the magnitude frequency response due to the pole is highly sensitive to the frequency selective at the steady state. An all-pass system composed of the poles and zeros, in principle, solves the problem of the frequency dependence of reverberation because the all-pass system has a flat magnitude frequency response at the steady state independent of the frequency. The frequency responses in concert halls or acoustic spaces, however, are not always flat from a practical perspective. A preferable reverberator would be still an interesting issue for audio engineering.

6.4.5 Decomposition of transfer function into product of minimum-phase and all-pass components

Suppose a transfer function

$$H(z^{-1}) = \frac{1 - bz^{-1}}{1 - az^{-1}}, \tag{6.131}$$

where $|b| > 1$. A system that has a transfer function like the one shown in Eq. (6.131) is called a nonminimum-phase system because the zero $z_0 = b$ is located outside the unit circle. Take the inversion of the zero such that

$$\overline{z}_0 = \overline{b} = b^{-1*}. \tag{6.132}$$

Substituting the inversion for z_0, then

$$\overline{H}(z^{-1}) = \frac{1 - b^{*-1}z^{-1}}{1 - az^{-1}}, \tag{6.133}$$

which is of minimum phase. In addition the magnitude on the unit circle where $z = e^{i\Omega}$,

$$\left|\overline{H}(e^{-i\Omega})\right| = \left|-b^{*-1}e^{-i\Omega}\frac{1 - b^{*}e^{i\Omega}}{1 - ae^{-i\Omega}}\right| = \left|-b^{*-1}\right|\left|H(e^{-i\Omega})\right|, \tag{6.134}$$

which indicates the magnitude frequency responses of $\overline{H}(e^{-i\Omega})$ and $H(e^{-i\Omega})$, are identical to each other within the scaling constant.

Changing the zero z_0 into its inversion means altering the nonminimum (minimum) phase into the minimum (nonminimum) phase system, keeping the magnitude frequency response within the scaling factor. Dividing the transfer function $H(z^{-1})$ by $\overline{H}(z^{-1})$ yields

$$G(z^{-1}) = \frac{H(z^{-1})}{\overline{H}(z^{-1})} = \frac{1 - bz^{-1}}{1 - b^{*-1}z^{-1}}, \tag{6.135}$$

which is identical to the all-pass system given by Eq. (6.128) where the pole and zero make an inversion pair. The magnitude frequency response is independent of the angular frequency Ω.

The definition of $G(z^{-1})$ in Eq. (6.135) leads to a formulation such that

$$H(z^{-1}) = \overline{H}(z^{-1}) \cdot G(z^{-1}), \tag{6.136}$$

which presents the transfer function $H(z^{-1})$ that can be decomposed into the product of the minimum-phase and all-pass components. The decomposition is described in the reference [11] from a signal processing standpoint by using the cepstral decomposition.

Audio equalization can be interpreted as a process of controlling the magnitude response of the minimum-phase components. In principle, the magnitude response approaches to flatness, similar to an all-pass system at the limit of the equalization. The reverberation due to the impulse response of the all-pass component might be prominent in the time domain as the equalization process goes on in the frequency plane.

6.5 Exercises

1. Derive the formula of the mean free path given in Eq. (6.11)

2. Confirm Eq. (6.30) through Eq. (6.22).

3. The three types of reverberation time formulae shown in Eq. (6.90) correspond to those for three-, two-, and one-dimensional reverberation fields. Derive the reverberation time formulas for two- and one-dimensional fields according to the sound intensity decay curve given by Eq. (6.20) for three-dimensional fields [2][18].

4. Show the relationship between the room impulse response given by Eq. (6.36) following the mirror-image theory and the frequency characteristics at the steady state given by Eq. (6.44) according to modal theory [2][6][8]. This is a complicated issue. The theoretical relationship between the geometrical acoustics represented by mirror-image theory and wave theory is one of the main issues since reference [8]. Eq. (6.88) and the three types of reverberation time formulae are examples of the hybrid formulae of geometrical and wave theory approaches [7][16].

5. Give an example in which a zero located far away from the unit circle could be prominent, given by the exponential windowing described by Eqs. (6.111) and (6.112).

6. Show the transfer function that is composed of an inversion pair of zeros with two poles at the origin in z−plane. Explain why the frequency response for the transfer function is called the linear phase. Transfer functions for all-pass systems are composed of inversion pairs of poles and zeros, while that for a linear phase system is made of inversion pairs of zeros.

7. Confirm Eq. (6.130) for all-pass systems.

8. Explain the terminologies listed below:
(1) density of mirror-image sources (2) mean free path (3) number of collisions for reflected sounds
(4) sound intensity (5) sound absorption coefficient (6) reverberation time
(7) equivalent sound absorption area (8) intensity decay curve (9) white noise
(10) steady-state response (11) eigenfrequency (12) modal function
(13) transfer function (14) zero (pole) of transfer function (15) modal density
(16) oblique, tangential, axial modes (17) modal bandwidth (18) modal overlap
(19) all pass (20) minimum phase (21) accumulated phase
(22) exponential windowing (23) inverse system (24) inversion of complex number
(25) nonminimum phase

References

[1] M. Tohyama, T. Koike, Fundamentals of Acoustic Signal Processing, Academic Press, 1998.
[2] M. Tohyama, Sound and Signals, Springer, 2011.
[3] L. Batchelder, Reciprocal of the mean free path, J. Acoust. Soc. Am. 36 (3) (1964) 551–555.
[4] M.R. Schroeder, New method of measuring of reverberation time, J. Acoust. Soc. Am. 37 (3) (1965) 409–412.
[5] M. Tohyama, Equivalent sound absorption area in a rectangular reverberant room (Sabine's sound absorption factor), J. Sound Vib. 108 (2) (1986) 339–343.
[6] J.B. Allen, D.A. Berkley, Image method for efficiently simulating small-room acoustics, J. Acoust. Soc. Am. 65 (4) (1979) 943–950.
[7] Y. Hirata, Geometrical acoustics for rectangular rooms, Acustica 43 (4) (1979) 247–252.
[8] P.M. Morse, R.H. Bolt, Sound waves in rooms, Rev. Mod. Phys. 16 (1944) 69–150.
[9] R.H. Lyon, Progressive phase trends in multi-degree-of-freedom systems, J. Acoust. Soc. Am. 73 (4) (1983) 1223–1228.
[10] R.H. Lyon, Range and frequency dependence of transfer function phase, J. Acoust. Soc. Am. 76 (5) (1984) 1435–1437.
[11] M. Tohyama, Waveform Analysis of Sound, Springer, 2015.
[12] M. Tohyama, A. Imai, H. Tachibana, The relative variance in sound power measurement using reverberation rooms, J. Sound Vib. 128 (1) (1989) 57–69.
[13] J. Davy, The relative variance of the transmission function of a reverberation room, J. Sound Vib. 77 (1981) 455–479.
[14] M. Tohyama, A. Suzuki, Space variances in the mean-square pressure at the boundaries of a rectangular reverberant room, J. Acoust. Soc. Am. 80 (3) (1986) 828–832.
[15] J. Blauert, N. Xiang, Acoustics for Engineers, Springer, 2008.
[16] M. Tohyama, S. Yoshikawa, Approximate formula of the averaged sound energy decay curve in a rectangular reverberant room, J. Acoust. Soc. Am. 70 (6) (1981) 1674–1678.
[17] Y. Hirata, Dependence of the curvature of sound decay curves and absorption distribution on room shapes, J. Sound Vib. 84 (1982) 509–517.
[18] M. Tohyama, Sound in the Time Domain, Springer, 2017.
[19] M.R. Schroeder, D. Hackman, Iterative calculation of reverberation time, Acta Acust. 45 (4) (1980) 269–273.
[20] M. Tohyama, A. Suzuki, Reverberation time in an almost-two-dimensional diffuse field, J. Sound Vib. 111 (3) (1986) 391–398.

[21] M.R. Schroeder, Improved quasi-stereophony and colorless artificial reverberation, J. Acoust. Soc. Am. 33 (1961) 1061–1064.
[22] Y. Ando, Signal Processing in Auditory Neuroscience; Temporal and Spatial Features of Sound and Speech, Elsevier, 2019.
[23] S. Lang, Complex Analysis, Springer New-York, Inc., 1999.
[24] M. Tohyama, R.H. Lyon, T. Koike, Reverberant phase in a room and zeros in the complex frequency plane, J. Acoust. Soc. Am. 89 (4) (1991) 1701–1707.
[25] M. Tohyama, R.H. Lyon, T. Koike, Phase variabilities and zeros in reverberant transfer function, J. Acoust. Soc. Am. 95 (1) (1994) 286–296.
[26] M. Tohyama, H. Suzuki, Y. Ando, The Nature and Technology of Acoustic Space, Academic Press, London, 1995.

CHAPTER

Intelligibility and reverberation

7

CONTENTS

7.1 Superposition of direct sound and reverberation

7.1.1 Ratio of direct sound and reverberation

Suppose that a receiving position is located at r (m) from a point source. The sound intensity at the receiving position is given by [1][2][3]

$$I_D = \frac{P_0}{4\pi r^2} \qquad (\mathrm{W/m^2}) \tag{7.1}$$

for the direct sound from the point source, where P_0 (W) gives the sound power output of the source. On the other hand, the sound intensity for the reverberation is estimated by [1][2][3]

$$I_R = \frac{4P_0}{\alpha S} = \frac{4P_0}{A}, \qquad (\mathrm{W/m^2}) \tag{7.2}$$

Acoustic Signals and Hearing. https://doi.org/10.1016/B978-0-12-816391-7.00015-2

according to Eq. (6.19), and assuming $-\log(1-\alpha)S \cong \alpha S$. The intensity I_R given by Eq. (7.2) is independent of the distance from the source, where α denotes the averaged sound absorption coefficient for the surrounding wall, S (m^2) shows the area of the wall, and $A = \alpha S$ (m^2) is called the equivalent absorption area. Taking the ratio of the direct to the reverberant energy (or intensity)

$$K = \frac{I_D}{I_R} = \frac{P_0}{4\pi r^2} / \frac{4P_0}{\alpha S} = \frac{\alpha S}{16\pi r^2} \quad (7.3)$$

is obtained. The ratio K in Eq. (7.3) is a function of the distance r (m) from the source. The energy ratio decreases in inverse proportion to r^2 as the distance increases. The distance r_c (m), where the ratio becomes unity as

$$K_c = \frac{\alpha S}{16\pi r_c^2} = 1 \quad \text{or} \quad r_c^2 = \frac{\alpha S}{16\pi} \quad (\mathrm{m}^2), \quad (7.4)$$

is called the critical distance, and is a representative parameter of room acoustics.

7.1.2 Intelligibility and distance from sound source under reverberant condition

Fig. 7.1 illustrates speech intelligibility for monosyllables (Japanese and Dutch) in a rectangular reverberant space [4] where the room volume is 640 (m^3) and the reverberation time is shown in the right panel in the figure [5].

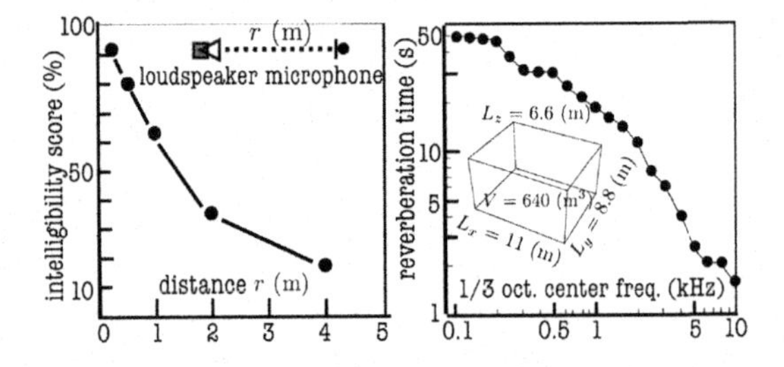

FIGURE 7.1

Left: Intelligibility for Japanese syllables in a rectangular reverberant room from Fig. 4 [4]. Right: Reverberation time in the reverberant room where intelligibility tests are performed from Fig. 2 [5].

The details of the intelligibility test conditions are described in reference [4]. The intelligibility decreases as the distance from the talker increases in the reverberant space. This result suggests that the intelligibility depends on the ratio of direct to reverberant sound energy in the reverberant space.

Experiments on measuring the articulation loss of syllables use carrier phrases before the target, as shown in Fig. 7.2 [6]. Reverberation of the carrier phrase reduces intelligibility (increases the articulation loss) under reverberant conditions. The reverberant energy of the carrier phrase decreases as the time interval ΔT increases between the carrier phrase and the target syllable. Suppose a squared impulse response $h^2(t)$ that specifies the reverberant condition such that

$$h_r^2(t) = a(r)\delta(t) + e^{-13.8t/T_R}, \tag{7.5}$$

where the squared impulse response is a superposition of the direct sound represented by $\delta(t)$ and reverberation expressed by the exponential function. The intensity of the direct sound can be changed by $a(r)$ depending on the sound source distance r (m), while the reverberation is an exponentially decaying function of time independent of the distance, T_R (s) denotes the reverberation time, and $-13.8 \cong \log_e 10^{-6}$ following Eq. (6.17).

Test syllables can be synthesized by convolution of the original syllables and impulse response assuming that $h_r(t) = \sqrt{h_r^2(t)} \cdot h_c(t)$, where $h_c(t)$ denotes a random sequence [6]. Articulation loss of the syllables is measured by setting the reverberation time to 3 (s) and changing the sound source distances by r (m) and ΔT (ms), as shown in Fig. 7.2. The results are plotted as shown in Fig. 7.2 [6][7] against the ratio as

$$R(r, \Delta T) = \frac{\int_0^{20(\mathrm{ms})} h_r^2(t)dt}{\int_{\Delta T}^{\infty} h_r^2(t)dt}. \tag{7.6}$$

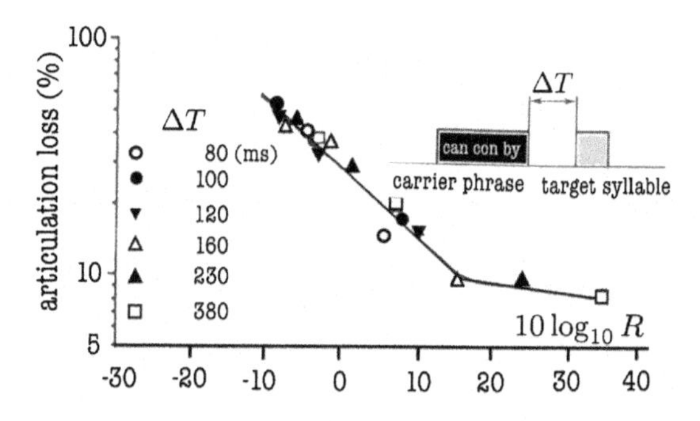

FIGURE 7.2

Articulation loss for syllables under reverberant conditions by setting $T_R = 3$ s and $r = 1, 2, 3, 4, 6, 8$ (m) from Figs. 2 and 5 [6].

The ratio in Eq. (7.6) indicates the energy ratio between the direct and initial part (within 20 (ms) of the target at the distance r and reverberation of the carrier phrase. The articulation loss rapidly decreases as the ratio increases. The energy ratio R recalls D_{30}, which defines the direct sound including early reflections within 30 (ms) after the physical direct sound. The subjective energy ratio D_{30} is popular in room acoustics [8][9], as is the physical energy ratio given by Eq. (7.3).

7.2 Initial decay curve of reverberation and intelligibility

7.2.1 Initial portion of reverberation decay and sound source distance

Intelligibility decreases as the distance from a source increases because the energy ratio becomes lower as the distance increases in a reverberant space. The effect of the sound source distance can also be seen on the initial decay rate of the reverberation decay curve. Fig. 7.3 displays the initial decay portion of the reverberation decay curves that are measured in a reverberant room whose volume is 183 (m^3) [10][11].

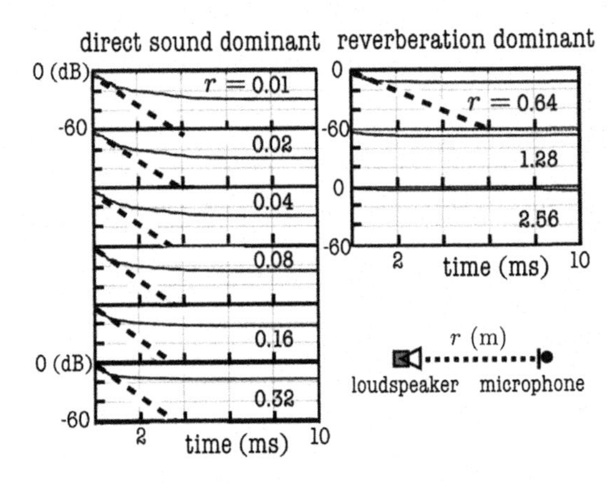

FIGURE 7.3

Samples of initial decay portions of reverberant energy decay curves in a room from Fig. 4.18 [10].

The impulse responses are exponentially windowed (see Eq. (6.111)) so that the reverberation time might be around 1 (s) [10][11]. Reverberation decay curves can be defined independently of the locations of the sound source and receiving positions, according to the room reverberation theory described in Chapter 6. However, the decay curves are rather sensitive to the sound source distance. The initial decay rate is higher, but is not highly sensitive to the sound source distance, in the region where the distance is smaller or the direct sound can be dominant at the receiving position. In contrast, the decay becomes remarkably slower as the receiving position is outside the direct-sound-dominant region. This change in the initial decay rate may correspond to the energy ratio of the direct sound and reverberation as a function of the sound source distance. Recalling that intelligibility depends on the sound source distance, intelligibility might increase as the initial decay rate increases, even in reverberant space.

7.2.2 Nonexponential decay curve and intelligibility

Initial decay rates can vary in rooms independently of the sound source distance. Nonexponential decay curves are observed in a rectangular room, as described in

Chapter 6, even after spatial averaging sound source and receiving positions in the room. It may be prominent if the floor, for example, is covered by absorbing materials. Rapid decay rates can be observed for the initial portion in the nonexponential decay curves because the initial part is mostly composed of the three-dimensional (oblique) waves, while the later part of the decay curve is mostly represented by two-dimensional (tangential) or one-dimensional (axial) wave modes. Rapid initial decay rates can help enhance the intelligibility, even in a reverberant space [4][12].

The right panel of Fig. 7.4 is an example of nonexponential decay curve [4][13][14].

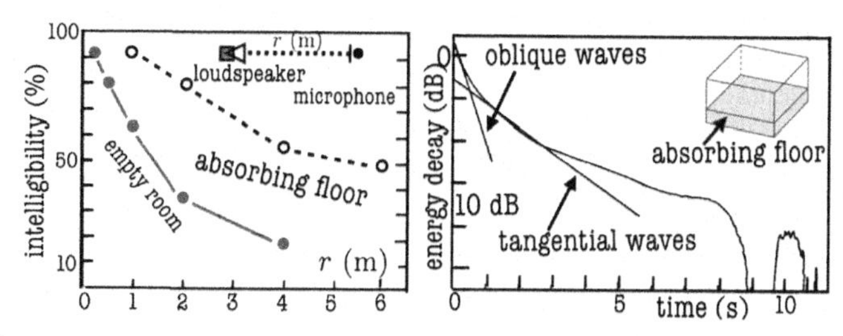

FIGURE 7.4

Right: Nonexponential decay curve in a rectangular reverberant room with absorbing floor from Fig. 2 [4]. Left: Intelligibility scores under reverberant conditions from Fig. 4 [4].

The left panel in Fig. 7.4 shows the intelligibility scores measured under reverberant conditions. The intelligibility tests were carried out in the same manner as for Fig. 7.1 using 100 Japanese Consonant-Vowel (CV) monosyllables and 50 Dutch Consonant-Vowel-Consonant Phoneme Balance words. The detailed conditions are described in reference [4][13]. The data in Fig. 7.1 are displayed once more for reference.

The speech intelligibility scores under the absorbing floor are much higher than those in the empty room, even under long distance conditions. The rapid initial decay due to the oblique waves may explain the higher intelligibility scores even under long reverberation in the mostly tangential reverberation field. Nonexponential reverberation decay with rapid initial decay under long reverberation may be a possibility in an auditorium where highly intelligible speech is combined with a rich reverberation of music.

7.3 Multi-array loudspeaker sources and intelligibility in reverberant space

7.3.1 Energy ratio of direct sound and reverberation for a pair of loudspeakers

Increasing the energy ratio of the direct sound to the reverberation may be helpful in enhancing speech intelligibility. Take the example of the energy ratio I_d/I_R for a single source in reverberant space once more. The ratio is given by

$$K_1 = \frac{\alpha S}{16\pi r^2} = \frac{A}{16\pi r^2}, \tag{7.7}$$

following Eq. (7.3), where r (m) denotes the distance between the source and microphone, and $A = \alpha S$ (m^2) is called the equivalent sound absorption area. The energy ratio can be extended into cases for multiple sources [15]. Suppose a pair of identical source types excited by the same signal. Assuming the distance between the pair of sources is greater than the half-wave length of the source frequency, f_c (Hz), then the reverberant sound rendered by the pair of sources can be uncorrelated to each other according to the spatial nature of random sound field (see subsection 9.3.1) [3]. The energy ratio becomes

$$K_2 = 2K_1 \tag{7.8}$$

because the energy of the direct sound is four times higher than that for the single source, while the energy of the reverberation is two times higher than that for the single source.

7.3.2 Energy ratio for loudspeakers in a circle

A pair of loudspeakers is extended to a loudspeaker array [15]. The left panel of Fig. 7.5 shows an array of sound sources in a circle of radius r.

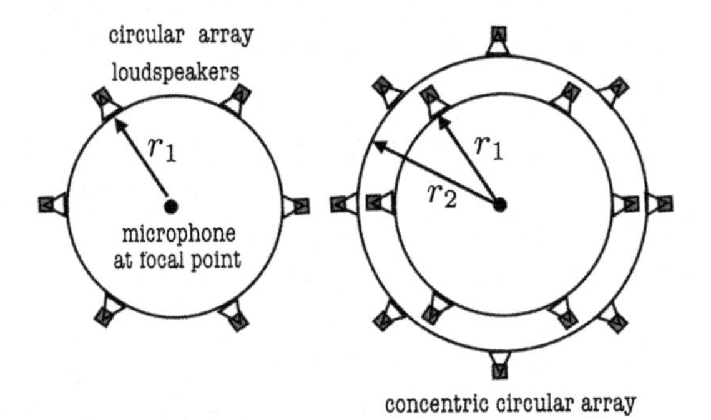

FIGURE 7.5

Left: Circular array of sound sources from Fig. 1 [15]. Right: Similarly concentric circular arrays from Fig. 2 [15].

The maximum number of uncorrelated sources for the reverberation is

$$N_{uc} = \frac{2\pi r}{\lambda_c/2} = \frac{4\pi r f_c}{c}, \tag{7.9}$$

where λ_c (m) gives the wavelength of the source frequency f_c (Hz) and c (m/s) is the speed of sound. The possible number of uncorrelated sources, N_{uc}, depends on the radius of the circle. Thus, the energy ratio for the circular array

$$K_{cr} = N_{uc} K_1 = \frac{A f_c}{4rc} \tag{7.10}$$

is derived.

The energy ratio for a spherical array can be straightforwardly derived such that

$$K_{sp} = N_{us} K_1 = \frac{A f_c^2}{c^2} \quad \text{and} \quad N_{us} = \frac{16\pi r^2}{c^2} f_c^2, \tag{7.11}$$

where K_1 is given by Eq. (7.7). The ratio is independent of r; however, it is still limited by the reverberant condition of the space, for example the equivalent absorption area A (m^2).

7.3.3 Concentric circular arrays

The right panel of Fig. 7.5 shows an example of an array of sources in two concentric circles of radii r_1 and r_2, where $r_2 - r_1$ is larger than the half-wave length for f_c [15]. Suppose that the maximum number of uncorrelated sources are located on the inner r_1 and outer r_2 circles, where N_{uc} is given by Eq. (7.9). Applying delayed signals by $\tau = (r_2 - r_1)/c$ (s) to the sound sources on the inner circle, the ratio at the center (focal point) is given by

$$K_{dcr} = \frac{(1+a)^2}{1+a^2 b} K_{cr}, \tag{7.12}$$

where K_{cr} refers to the inner circle, P_1 (W) (or Q_1 (m^3/s)) and P_2 (or Q_2) respectively denote the sound power output (or the volume velocity) of the sources on the inner and outer circles, and

$$a = \frac{Q_2}{Q_1} = \sqrt{\frac{P_2}{P_1}} \quad \text{and} \quad b = \frac{r_2}{r_1}, \tag{7.13}$$

where $r_2 > r_1$. The energy ratio given by Eq. (7.12) is highest when

$$\text{Max}[K_{dcr}] = \left(1 + \frac{1}{b}\right) K_{cr} \quad \text{for } a = \frac{1}{b} < 1. \tag{7.14}$$

The power of the outer-layer sources must be lower (inversely proportional to a radius of the circle) than that for the inner layer so that the energy ratio can reach a maximum. The improvement or enhancing effect on the energy ratio by adding the outer layer is

$$\frac{K_{dcr}}{K_{cr}} < 1 + \frac{1}{b} < 2, \tag{7.15}$$

which decreases as the radius of the outer circle increases.

Enhancing the direct sound in reverberant space would also be possible by constructing a linear loudspeaker array [15] instead of a circular array. Similarly, microphone arrays are possible in addition to loudspeaker arrays. Intelligibility tests have confirmed the enhancement by loudspeaker or microphone arrays, as described in the references [16][17]. The results in the references suggest that the intelligibility in the reverberant space depends basically on the energy ratio. The intelligibility can be high within the critical distance where the direct sound is prominent, even in the reverberant space. The loudspeaker (or microphone) arrays work so that the critical distance might be virtually longer by increasing the energy ratio at the focal point in the reverberant space.

7.4 Modulation envelope and reverberation condition

7.4.1 Reverberation effects on envelopes

Reverberation changes envelopes of transmission signals through reverberant space. The left panel of Fig. 7.6 shows a waveform of intelligible speech in an anechoic room with its power spectrum.

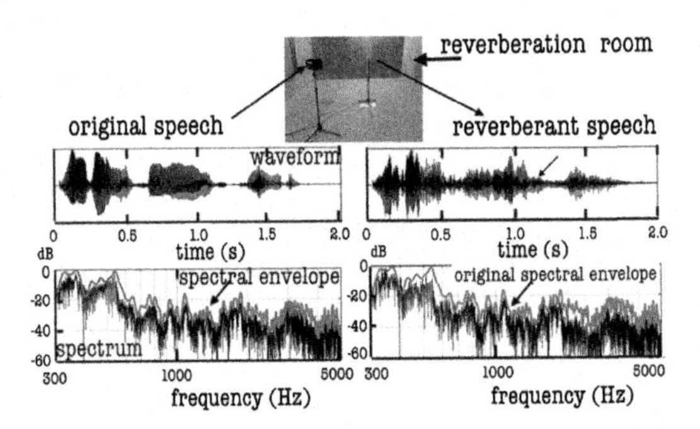

FIGURE 7.6

Waveforms and power spectral behavior of speech samples in an anechoic room (left) and in a reverberant room (right) from Figs. 4.2 and 4.3 [10].

Similarly the right panel of Fig. 7.6 illustrates the reverberant waveform and its power spectrum. The envelopes of the waveforms (left and right upper panels) in the time domain seem clearly different from each other; however, interestingly, the spectral envelope may be mostly preserved even under reverberant conditions. The deformation in the envelopes of the waveforms may lose the intelligibility of transmitting speech through reverberant space. In contrast, the spectral envelope may keep some source information even in reverberant space. In addition, Fig. 7.6 implies narrowband envelopes are the keys to preserving intelligibility instead of the whole-band envelopes.

7.4.2 Transmission of envelope in reverberant space

Suppose a modulated signal

$$x(t) = e(t) \cdot n(t), \tag{7.16}$$

where $n(t)$ denotes white noise. The transmitted signal $y(t)$ in a reverberant space is formulated

$$y(t) = x * h(t), \tag{7.17}$$

where $h(t)$ denotes the impulse response from the source position to the receiving position in the reverberant space. Taking the ensemble average of the squared receiving signal $y^2(t)$, then

$$\mathrm{E}[y^2(t)] = N \int_0^t e^2(t-\tau)h^2(\tau)d\tau, \tag{7.18}$$

where

$$\mathrm{E}[n^2(t)] = N \tag{7.19}$$

because of the randomness of white noise [18].

Eq. (7.18) indicates that the squared envelope of the receiving signal can be expressed as the convolution of the squared envelope and the squared impulse response of the reverberant path subject to that the carrier $n(t)$ is white noise. Taking the Fourier transform, the squared-envelope spectrum is represented by

$$X_2(\omega) \cdot H_2(\omega) = Y_2(\omega), \tag{7.20}$$

where $X_2(\omega)$ and $H_2(\omega)$ are the Fourier transforms of the squared envelope and the squared impulse response, respectively. The formulation in Eq. (7.20) corresponds to the squared version for the relationship between input and output responses in linear systems.

The deformation of the envelope in the time domain can be formulated by taking an example of modulated envelope:

$$e^2(t) = \frac{1 + \cos\omega_e t}{2}. \tag{7.21}$$

Assuming the carrier to be white noise, the squared envelope becomes

$$e_{rev}^2(t) = \frac{1 + m\cos(\omega_e t + \phi(\omega_e))}{2} \tag{7.22}$$

under reverberant conditions, according to Eq. (7.18), where

$$m = \frac{|H_2(\omega)|_{\omega=\omega_e}}{|H_2(\omega)|_{\omega=0}} = |H_2(\omega)|_{\omega=\omega_e} \tag{7.23}$$

by normalizing $|H_2(\omega)|_{\omega=0} = 1$. Here m is called the modulation index, and the Fourier transform

$$H_2(\omega) = \int_0^{\infty} h^2(t)e^{-i\omega t}dt = |H_2(\omega)|e^{i\phi(\omega)} \tag{7.24}$$

is called the (complex) modulation transfer function [18][19][20]. The modulation index $m(\omega_e)$ indicates the magnitude spectral characteristics of the squared envelope and $\phi(\omega_e)$ is the phase spectral property of the squared envelope.

7.4.3 Spectral deformation of squared envelope by reflections

As implied by D_{30} in Section 7.1.2, early reflected sounds can be helpful to intelligibility. The relationship given in Eq. (7.18) might partly explain the positive effect of early reflections on intelligibility in terms of the envelope. Suppose a conceptual model for impulse response representing the distribution (or density) of early reflections such that

$$h^2(t) = \begin{cases} 1/T & -T/2 < t < T/2 \\ 0 & \text{others} \end{cases}, \tag{7.25}$$

where the origin of the time is shifted for simplicity. Taking the Fourier transform of the equation above,

$$H_2(\omega) = \frac{1}{T}\int_{-T/2}^{T/2} e^{-i\omega t}dt = \frac{\sin \omega T/2}{\omega T/2} \tag{7.26}$$

is obtained. The function given by Eq. (7.26) is called the sinc function.

According to D_{30} [8][9], set $T = 30$ (ms). The sinc function above becomes 0 at the frequency such that

$$F_c = 1000/30 \cong 33. \qquad \text{(Hz)} \tag{7.27}$$

The frequency F_c implies that the spectrum of the squared envelope is not severely deformed as long as $f < F_C \cong 33$ (Hz). The spectral properties of squared envelopes of intelligible speech samples may meet the frequency range [19] [20]. The motivation of D_{30} would be partly understood in terms of the spectral property of squared envelopes of intelligible speech samples.

7.5 Modulation transfer functions and linear systems

7.5.1 Modulation index and sound source distance

Modulation indexes or modulation transfer functions depend on the sound source distance or the energy ratio of the direct sound and reverberation in acoustic space. The effect of reverberation is represented by the modulation index. The reverberation makes the modulation envelope shallower in the time domain. The intelligibility is sensitive to the modulation index in the reverberant space. Compare Figs. 7.7 and the left panel of Fig. 7.1; the speech intelligibility can be estimated by the modulation index or the modulation transfer function.

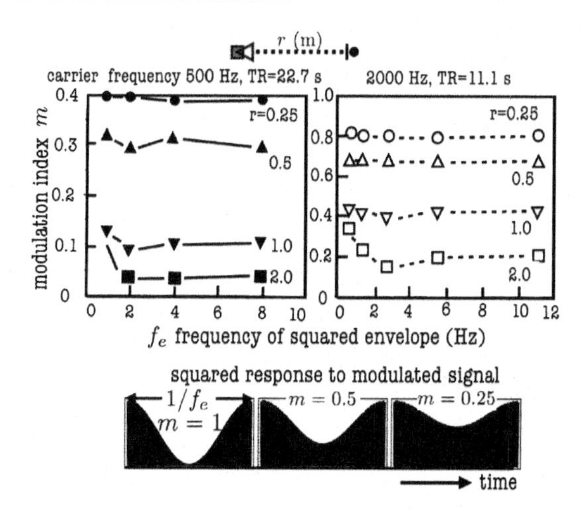

FIGURE 7.7

Modulation index and sound source distance in reverberant space from Fig. 3.7.2 [13].

The estimators on speech intelligibility such as RASTI or MTF-STI are developed in detail in the references [19] and [20] based on the modulation index. The modulation might be almost perfect within a small sound source distance even in a reverberant space. Improvement of speech intelligibility in Fig. 7.4 or an increase in the energy ratio by the loudspeaker array in Fig. 7.5 can be confirmed by RASTI in the references [4][15].

7.5.2 Spectral deformation of envelope by sound source distance

Fig. 7.8 shows an example of D_{30} in a reverberant space as a function of the sound source distance [10][11]. The subjective energy ratio D_{30} can be preserved as long as the sound source distance is within the direct-sound dominant region. Recalling Eq. (7.27), the spectral property of squared envelopes for an intelligible speech record can be preserved for spectral components lower than 20 (Hz). Fig. 7.9 displays examples of the spectral behavior for 1/4-octave-band squared envelopes of speech records under reverberant conditions in which D_{30} is measured, as shown by Fig. 7.8 [11][21]. The envelope spectrum mostly ranges to frequencies lower than 20 (Hz), and the prominent frequency bands are 297 (Hz) and 707 (Hz) for 1/4-octave frequency bands, respectively in Fig. 7.9. The lower right part of the figure shows the envelope energy in the two prominent frequency bands as a function of the sound source distance in reverberant space in which D_{30} is illustrated in Fig. 7.8. The envelope energy is kept as long as the D_{30} is preserved (in the direct-sound dominant

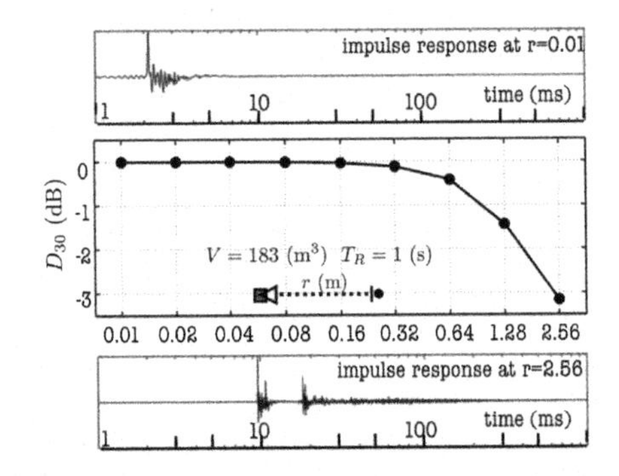

FIGURE 7.8

Example of D_{30} as a function of sound source distance in a reverberant space with samples of impulse response records from Fig. 5.4 [10].

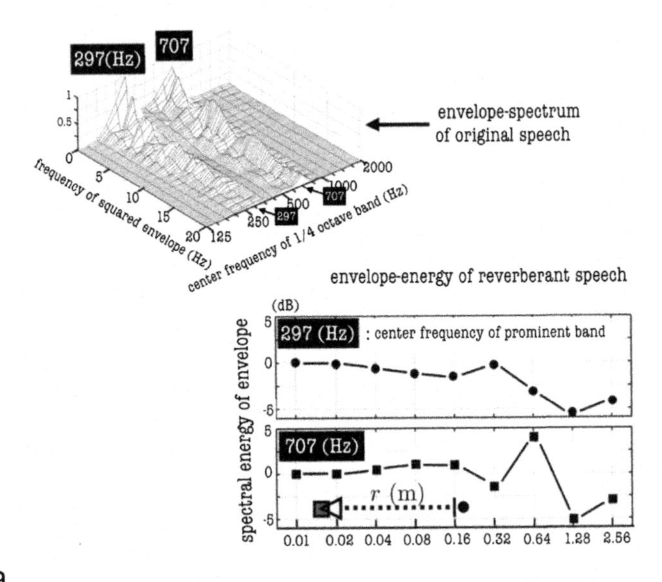

FIGURE 7.9

Example of spectral behavior for 1/4-octave-band squared envelopes for speech sample and envelope energy in prominent frequency bands in reverberant space as a function of sound source distance in which D_{30} is measured as shown in Fig. 7.8 from Figs. 3 and 4 [21].

region) even in the reverberant space. In contrast, the envelope energy can be deformed from the original one if the sound source distance exceeds the direct-sound dominant region in which D_{30} rapidly decreases. The deformation of the envelope makes speech samples unintelligible. Recalling that the lowest audible frequency range of the hearing organ is around 20 (Hz), the envelope composed of the frequencies lower than 20 (Hz) might be sensed in the time domain rather than the frequency domain. The speech intelligibility conveyed by the narrowband envelopes would be a typical auditory event in the time domain.

7.5.3 Complex modulation transfer function in linear systems

The modulation transfer function is defined by Eq. (7.24). Recalling the Fourier transform of $x_w(n) = w(n)x(n)$ such that [2]

$$X_W(\Omega) = X * W(\Omega), \tag{7.28}$$

the modulation transfer function $H_2(\omega)$ can be rewritten as

$$H_2(\omega) = H * H(\omega), \tag{7.29}$$

where $H(\omega)$ denotes the Fourier transform of the impulse response $h(t)$. Suppose that the impulse response is of the linear phase:

$$H(\omega) = |H(\omega)|\mathrm{e}^{-\mathrm{i}\omega\tau}. \tag{7.30}$$

The modulation transfer function can be rewritten as [13]

$$H_2(\omega) = \mathrm{e}^{-\mathrm{i}\omega\tau}\int |H(\omega')H(\omega - \omega')|d\omega'. \tag{7.31}$$

Interestingly, the squared response $h^2(t)$ is again of the linear phase when the impulse response $h(t)$ is of the linear phase. A squared envelope such as

$$e^2(t) = \frac{1 + \cos\omega t}{2} \tag{7.32}$$

becomes

$$e^2_{Lph}(t) = \frac{1 + m(\omega)\cos\omega(t - \tau)}{2} \tag{7.33}$$

after the linear-phase system, where $m(\omega)$ is determined by the magnitude of the complex modulation transfer function and the delay τ is independent of the modulation frequency because of the linear phase [18], even if the modulation frequency changes. The deformation for the envelope due to linear-phase systems can be represented by the modulation index.

In contrast, suppose that the linear system is of the all-pass system:

$$H(\omega) = A\mathrm{e}^{\mathrm{i}\theta(\omega)}. \tag{7.34}$$

The modulation transfer function becomes

$$H_2(\omega) = A^2\int \mathrm{e}^{\mathrm{i}\theta(\omega')}\cdot\mathrm{e}^{-\mathrm{i}\theta(\omega'-\omega)}d\omega', \tag{7.35}$$

which is the phase correlation function, where the phase is assumed by an odd function of the frequency [3][22]. The micro structure of the phase largely changes the envelopes in the time domain.

The reverberation effect on the envelope due to the all-pass component, can be estimated by the complex modulation transfer function including the frequency-dependent time delay of the envelope. Fortunately, the phase-trend of the impulse response mostly follows a linear phase in the narrowband, for example a $1/3 - 1/4$ octave band under reverberant conditions, as will be mentioned in Section 8.1.3 [3][13][23][24]. The modulation index may be a possible measure for intelligibility in reverberant space.

7.6 Exercises

1. Derive the formula of the energy ratio of the direct sound and reverberation given in Eq. (7.3).

2. Confirm Eqs. (7.10) and (7.11).

3. Derive Eqs. (7.12) recalling that [1][2][3]

$$\begin{aligned} p &= \mathrm{i}\omega\rho\frac{Q}{4\pi r}\mathrm{e}^{\mathrm{i}\omega(t-r/c)} \quad \text{(Pa)} \\ v &= \frac{p}{\rho c} \quad \text{(m/s)} \\ I &= \frac{1}{2}\Re\overline{[p \cdot v^*]} = \frac{P_0}{4\pi r^2} \quad (\mathrm{W/m^2}) \\ P_0 &= \frac{\rho\omega^2 Q^2}{8\pi c} \quad \text{(W)}, \end{aligned} \tag{7.36}$$

where p and v denote the sound pressure and particle velocity at a position separated from a point source by r (m), I and P_0 give the intensity at the receiving position and the sound power output of the point source, Q ($\mathrm{m^3/s}$) shows the volume velocity (or the strength) of the source, ρ ($\mathrm{kg/m^3}$) gives the volume density of the medium, and $\overline{(*)}$ is taking the average in a single period.

4. Observe the effects of reverberation on the macroscopic signatures of speech waveforms and their spectral properties in Fig. 7.6. The power spectral envelope of the original envelope can be mostly preserved even under reverberant condition, while the temporal envelope of the waveform may be largely deformed by the reverberation. It would be an intriguing question how the reverberation effects on the speech sample can be interpreted.

5. Formulate Eq. (7.22) following Eq. (7.18). Explain why the squared envelope can be expressed in the convolution.

6. Consider why early reflections within 30 (ms) implied by D_{30} may not be harmful to speech intelligibility in reverberant space.

7. Show an example, following the modulation transfer function, where the modulation index decreases as a function of the sound source distance in reverberant space.

8. Suppose a signal $y(n) = w(n)x(n)$, where $n = 0, 1, \ldots, N-1$. Show the $N-$point discrete Fourier transform of $y(n)$ such that $Y(k) = W * X(k)$, where $W(k)$ and $X(k)$ denote the $N-$point discrete Fourier transform of $w(n)$ and $x(n)$, respectively [2]. The same question is given in Exercise 4.5.5.

9. Confirm Eq. (7.33) according to Eq. (7.29).

10. Explain terminologies listed below:
(1) sound power output of a point source (2) intensity of direct sound (3) ratio of direct and reverberation intensity
(4) intensity of reverberation (5) equivalent sound absorption area (6) critical distance
(7) carrier phrase for intelligibility measurement (8) subjective ratio between direct energy and reverberant energy (9) initial decay rate of reverberation decay curve
(10) white noise (11) modulation index (12) modulation transfer function
(13) complex modulation transfer function (14) sinc function (15) linear phase
(16) all-pass system (17) oblique wave (18) tangential wave
(19) temporal envelope (20) spectral envelope

References

[1] J. Blauert, N. Xiang, Acoustics for Engineers, Springer, 2008.

[2] M. Tohyama, T. Koike, Fundamentals of Acoustic Signal Processing, Academic Press, 1998.

[3] M. Tohyama, Sound and Signals, Springer, 2011.

[4] H. Nomura, H. Miyata, T. Houtgast, Speech intelligibility and modulation transfer function in non-exponential decay fields, Acustica 69 (1989) 151–155.

[5] H. Miyata, H. Nomura, T. Houtgast, Speech intelligibility and subjective MTF under diotic and dichotic listening conditions in reverberant sound fields, Acustica 73 (1991) 200–207.

[6] H. Miyata, T. Houtgast, Weighted MTF for predicting speech intelligibility in reverberant sound fields, in: EUROSPEECH'91, Second European Conference on Speech Communication and Technology, Genova, Italy, September 24–26, 1991, pp. 289–292.

[7] M. Tohyama, Modern techniques for improving speech intelligibility in noisy environments, in: Michel Vallet (Ed.), Noise and Man'93, Proc. 6-th Int. Congress, Noise as a Public Health Problem, Vol. 3, Nice, France, 1993, pp. 238–246.

[8] R. Thiele, Richtungsverteilung und Zeitfolge der Schallrueckwuerfe in Raeumen, Acustica 3 (1953) 291–302.

[9] T.J. Schlutz, Acoustics of the concert hall, IEEE Spectr. (June 1965) 56–67.

[10] M. Tohyama, Waveform Analysis of Sound, Springer, 2015.

[11] Y. Hara, Sound perception and temporal-spectral characteristics in sound field near sound source, PhD Thesis No. 126, Kogakuin University, Tokyo, Japan, Jan. 27, 2014.

[12] B. Yegnanarayana, B. Ramakrishna, Intelligibility of speech under non-exponential decay conditions, J. Acoust. Soc. Am. 58 (4) (1975) 853–857.
[13] M. Tohyama, H. Suzuki, Y. Ando, The Nature and Technology of Acoustic Space, Academic Press, London, 1995.
[14] Y. Hirata, A method of eliminating noise in power responses, J. Sound Vib. 84 (1982) 593–595.
[15] H. Nomura, M. Tohyama, T. Houtgast, Loudspeaker arrays for improving speech intelligibility in a reverberant space, J. Audio Eng. Soc. 39 (5) (1991) 338–343.
[16] H. Nomura, H. Miyata, T. Houtgast, Microphone arrays for improving speech intelligibility in a reverberant or noisy space, J. Audio Eng. Soc. 41 (10) (1993) 771–781.
[17] H. Nomura, H. Miyata, T. Houtgast, Speech intelligibility enhancement by a linear loudspeaker array in a reverberant field, Acustica 77 (1993) 253–261.
[18] M.R. Schroeder, Modulation transfer functions: definition and measurement, Acustica 49 (3) (1981) 179–182.
[19] T. Houtgast, H.J.M. Steenecken, R. Plomp, Predicting speech intelligibility in rooms from the modulation transfer function, Acustica 46 (1980) 60–72.
[20] T. Houtgast, H.J.M. Steeneken, R. Plomp, A review of the MTF concept in room acoustics and its use for estimating speech intelligibility in auditoria, J. Acoust. Soc. Am. 77 (3) (1985) 1069–1077.
[21] Y. Hara, Y. Takahashi, K. Miyoshi, Narrow-band spectral energy analysis close to sound source (in Japanese), J. Inst. Electron. Inf. Comm. Eng. Jpn. J 97-A (3) (2014) 221–223.
[22] M. Kazama, S. Gotoh, M. Tohyama, T. Houtgast, On the significance of phase in the short term Fourier spectrum for speech intelligibility, J. Acoust. Soc. Am. 127 (3) (2010) 1432–1439.
[23] M. Tohyama, R.H. Lyon, T. Koike, Reverberant phase in a room and zeros in the complex frequency plane, J. Acoust. Soc. Am. 89 (4) (1991) 1701–1707.
[24] M. Tohyama, R.H. Lyon, T. Koike, Phase variabilities and zeros in reverberant transfer function, J. Acoust. Soc. Am. 95 (1) (1994) 286–296.

CHAPTER

Subjective evaluation for coherent region in reverberant space

8

CONTENTS

8.1 Coherent length from sound source in reverberant space

8.1.1 Wave traveling in free space

Sound waves from a spherical source travel as symmetrical spherical waves in three-dimensional free space [1][2][3]. The phase difference in the traveling wave between the source and observation points in the space is estimated by

$$\phi(r) = -k \cdot r, \qquad \text{(rad)} \tag{8.1}$$

where r (m) denotes the distance between the source and observation points, $k = 2\pi f/c$ (rad/m), c (m/s) is the speed of sound in the space, and f (Hz) gives the

Acoustic Signals and Hearing. https://doi.org/10.1016/B978-0-12-816391-7.00016-4

frequency of the traveling wave. This type of phase function of kr is called the propagation phase [4][5]. The propagation phase is a linear function of the frequency f. The phase that is proportional to the frequency is also called the linear phase. In a free space waves travel without the deformation of waveforms within the time delay. This type of delay without any deformation is called the pure delay.

The pure delay is linear phase; however, linear phase is not necessarily the pure delay. Linear phase can be made by the transfer function composed of inversion pairs of zeros, as mentioned in Exercise 6.5.6 [6][7]. The difference between the linear-phase and pure delay systems is the frequency dependence of the magnitude frequency responses. Examples of linear-phase systems composed of inversion pairs of zeros imply the possibility to realize a trend of the linear phase in a reverberant space where magnitude frequency responses can be frequency dependent [6][7].

8.1.2 Accumulated phase of traveling waves in two-dimensional reverberant space

As described in Section 2.5, the trend of the accumulated phase follows the propagation phase between a pair of source and the receiving positions taken in a finite one-dimensional vibration system such as string vibration. For two-dimensional systems, the reverberation phase can be observed rather than the propagation phase [2][3][4][5]. It is expected, however, that the direct wave propagates to some extent toward a receiving point from a source when the distance r (m) from the source is not so far [8]. The region in which the direct wave travels with the propagation phase of kr is called the coherent region or coherent length representing the range of the region [4][5][8].

As developed in Section 6.4.3, the accumulated phase can be estimated by the numbers of the poles and zeros of the transfer function from the source to a receiving position in the space. Suppose the frequency characteristics of the transfer function in a two-dimensional space such as a membrane:

$$H(\omega, \mathbf{r_s}, \mathbf{r_o}) = \sum_{l,m} \frac{F(\mathbf{r_s})F(\mathbf{r_o})}{\omega^2 - \omega_{lm}^2}. \tag{8.2}$$

Here

$$F(\mathbf{r_s}) = \sin\frac{l\pi x_s}{L_x} \sin\frac{l\pi y_s}{L_y} \quad \text{for } \omega_{lm} = c\sqrt{\left(\frac{l\pi}{L_x}\right)^2 + \left(\frac{m\pi}{L_y}\right)^2} \quad \text{(rad/s)} \tag{8.3}$$

and $\mathbf{r_s}(x_s, y_s)$ is the position of the point source, similarly $\mathbf{r_o}(x_o, y_o)$ that of the receiving position, and ω_{lm} is the eigen angular frequency (1/s), L_x (m) and L_y (m) are the lengths of the sides for the rectangular membrane, and c (m/s) gives the speed of sound for the vibration of the membrane [2][3]. The occurrences of zeros between two adjacent poles along the frequency axis depend on the relationship between the pair of residues, as mentioned in Section 5.5.1 [2][4][5][9]. The number of zeros can be estimated by using the probability in the residue sign change [4][5]. The residue signs depend on the distance between the source and receiving positions. Fig. 8.1 illustrates the probability that a pair of adjacent poles has opposite sign residues under the condition that the sound source and receiving positions are randomly sampled in the two-dimensional space [2].

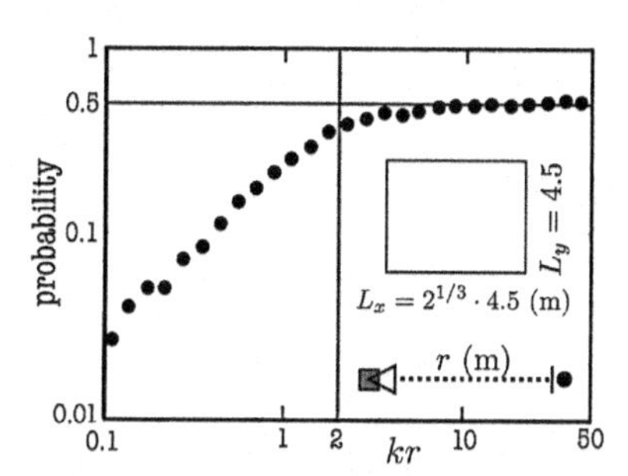

FIGURE 8.1

Probability in residue sign changes in two-dimensional space from Fig. 7.4.5 [2].

In the figure the probability is calculated for 300 randomly chosen source-and-receiving position pairs in the space where $L_x = 2^{1/3} \cdot 4.5$ (m), $L_y = 4.5$ (m).

The probability increases almost in linear proportion to kr. The probability, however, is about 1/2 when kr is greater than $kr \cong 2$ [3][4][5]. There is a single zero, between a pair of adjacent poles with same sign residues, on the pole-line that connects the adjacent poles, while no zeros occur on the pole-line when the adjacent poles have opposite sign residues to each other at the first approximation [4][5]. Entering into the details a little bit more, there can be three cases for the zero occurring between the adjacent pair of poles under the opposite-sign residues as shown by Fig. 8.2 [2][3][4][5][9].

The occurrences of the zeros are intensively investigated in the references [2][3][4][5][9] for the following frequency response:

$$H(\omega) = \frac{A}{\omega - \omega_1} + \frac{B}{\omega - \omega_2} + R(\omega). \tag{8.4}$$

Here $R(\omega)$ gives the remainder function that represents the sum of the contribution from poles excluding the two adjacent poles. The remainder function can be assumed as a slowly varying function that is almost constant in the interval between the adjacent pair of poles.

According to the relationship between the zero occurrence and residue signs, the number of zeros on the pole-line can be estimated by using a stochastic model in the residue sign change. Assume the transfer function based on the poles with their residue signs such that

$$H(\omega) = \sum_{n=1}^{N} \frac{A_n}{\omega - \omega_n}, \tag{8.5}$$

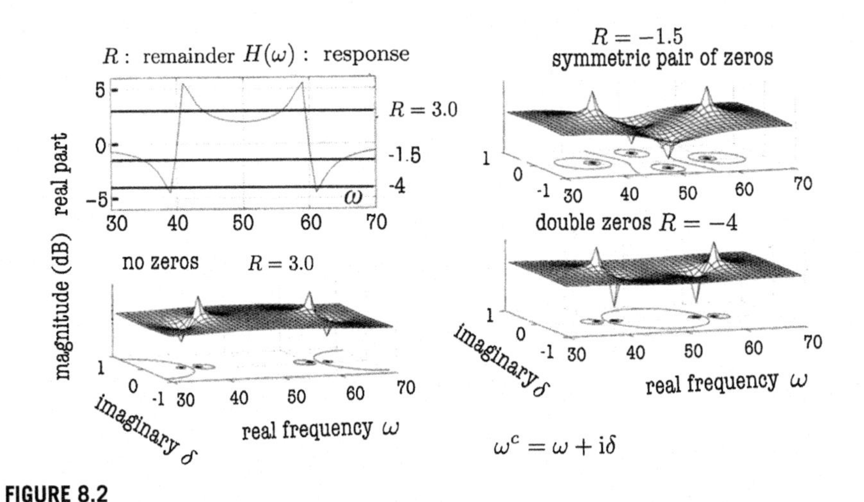

FIGURE 8.2

Examples of zeros in complex frequency plane under opposite-sign residues and remainders from Fig. 7.4.4 [2].

where $|A_n| = 1$, the probabilities of residue sign changes between the adjacent pair of poles may follow a binomial distribution [10], and the poles are distributed according to a Poisson distribution with the average of N_p [2][10][11]. Fig. 8.3 illustrates a distribution of zeros on the complex frequency plane.

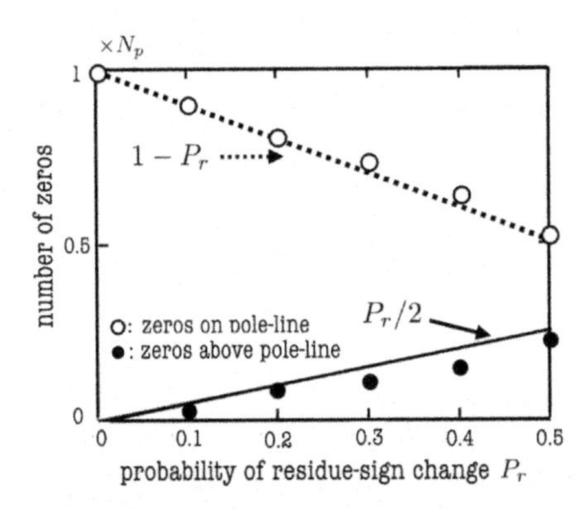

FIGURE 8.3

Distribution of zeros on a complex frequency plane for Eq. (8.5) taking 100 random samples from Fig. 7.4.6 [2].

The number of zeros on the pole-line decreases following

$$N_{z,\text{on-line}} \cong (1 - P_r)N_p, \tag{8.6}$$

while the off-line zeros (or above the pole-line on the complex frequency plane) increases as

$$N_{z,\text{ above-line}} \cong \frac{1}{2} P_r \cdot N_p = N_{z,\text{ below-line}}, \tag{8.7}$$

where N_p denotes the expectation of the number of poles and P_r gives the probability in the residue sign change.

The phase accumulation as the distance r from the source increases might be the process of decreasing the number of on-line zeros because the phase accumulation by the zeros on the lower and upper half-planes cancel each other due to the symmetric locations of the off-line zeros with respect to the pole-line or the real frequency axis [4][5][9]. The accumulated phase can be estimated as

$$\Phi(\omega) = -\pi(N_p - N_{z,\text{on-line}}) \cong -\pi P_r N_p \tag{8.8}$$

by using the number of poles and the probability in the residue sign changes. The phase is called reverberation phase for $P_r \cong 1/2$ that may exceed the propagation phase by kr.

8.1.3 Propagation and reverberation phase in three-dimensional space

In two-dimensional systems, the accumulated phase reaches the reverberation phase when kr exceeds around 2 or the probability in residue sign changes reaches around 1/2. The reverberation phase that can be estimated by

$$\Phi_{rev}(\omega) \cong -\pi P_r N_p = -\pi N_p/2, \tag{8.9}$$

which is mostly independent of the distance r, where $P_r = 1/2$. However, the propagation phase would still possibly be expected even in the reverberant space when the distance is not as far from the sound source. The probability in residue sign change smaller than 1/2 can be dependent on the distance in the region not far from the source, as shown in Fig. 8.1. Fig. 8.4 shows experimental results on the phase responses [12] in a reverberation room. The results imply transition of the phase responses from the propagation phase following kr to the reverberation phase for $kr > \pi$ in a reverberant space [3]. The reverberation phase estimated by $-\pi N_p/2$ shows the maximum possible phase delay under the condition that the damping effect is negligibly small compared with the eigenfrequencies. The eigenfrequency or the pole can be expressed as

$$\omega_N^c = \pm\omega_N + i\delta_N \text{ and } 2\delta_N \cdot T_R = \log_e 10^6 \cong 13.8 \tag{8.10}$$

in the complex frequency plane, where $\delta_N \cong 6.9/T_R$ gives the damping constant, T_R gives the reverberation time of the reverberant space, as described by Eq. (6.17) [2][3]. The pole-line is close to the real frequency axis when δ_N is sufficiently small. The symmetric pair of zeros with respect to the pole-line or the real frequency axis

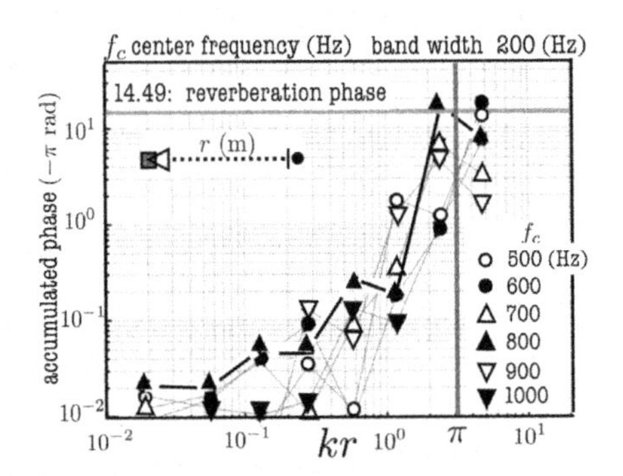

FIGURE 8.4

Sample of phase responses in reverberation room from Fig. 10 [12].

cancels out the phase accumulation; therefore, the reverberation phase can be estimated by Eq. (8.9). However, the reverberation phase can decrease as the damping factor increases. Increasing δ_N, the pole-line goes up to the real frequency axis [9]. This movement of the pole-line breaks down the symmetric locations of the off-line zero pairs with respect to the real frequency axis; the number of nonminimum-phase zeros that are left in the lower half-plane consequently decreases. This elimination in the number of nonminimum-phase zeros decreases the accumulated phase.

The details of the distribution of zeros below the pole-line are described in references [3][13][14][15][16][17]. According to the Cauchy distribution, which estimates the density of the zeros below the pole-line, the density of the nonminimum-phase zeros is given by

$$n_z^+ \cong \frac{1}{2} \cdot \frac{1}{2\pi\delta_0} \ \text{(s)} \ \text{and} \ \delta_0 \cong \frac{13.8}{2 \cdot T_R} \ \text{(1/s)} \tag{8.11}$$

under the high modal overlap condition, where δ_0 denotes the distance between the real frequency axis and the pole-line. Interestingly, the density of the nonminimum-phase zeros is independent of the frequency band. It indicates that the reverberation phase trend can be linear phase [3][13][14]. Assuming $T_R = 1$ (s), for example, the frequency bandwidth, and $\Delta\omega = 2\pi \cdot 200$ (rad/s), then the reverberant phase becomes [12]

$$\Phi(\omega) = -\pi n_z^+ \cdot \Delta\omega = -\pi \cdot \frac{100}{6.9} \cong -14.49\pi, \ \text{(rad)} \tag{8.12}$$

which corresponds to that in Fig. 8.4.

The number of nonminimum-phase zeros decreases as $T_R \to 0$ or $\delta_0 \to \infty$, even if the probability in the residue sign change reaches 1/2. Assuming $N_p = N_z^-$, where N_p denotes the number of poles and N_z^- is for the minimum-phase zeros in the limit case, the reverberant phase can be zero because of the cancelation of phase accumulation by the poles and minimum-phase zeros. However, only the propagation phase may remain, as suggested by the phase for a one-dimensional system described in Section 2.5.2. The time delay for the wave traveling from the source to receiving positions makes zeros at infinity below the real frequency axis instead of the minimum-phase zeros, and thus the zeros yield the propagation phase, even if no reverberation phase, such as a free field, can be seen in the space.

The zeros at the −infinity (below the real frequency axis) yields the phase trend of propagation, even in a reverberant space. Take the minimum-phase component from the response in a reverberant space. No phase accumulation can be seen for the minimum-phase component; however, zeros at +infinity in the upper half-plane are newly rendered by taking the minimum-phase component. The zeros at the −infinity yield the phase trend of propagation, and therefore removing the −infinity zeros removes the propagation phase trend. No propagation-phase trend in the whole of frequency range, however, implies that the trend may be left in the local frequency characteristics in the narrow frequency ranges.

Fig. 8.5 shows examples of the propagation phase trend in the reverberant space as shown in Fig. 8.4.

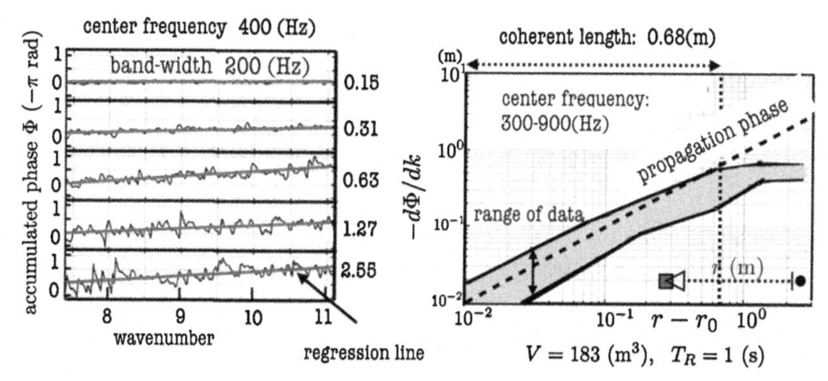

FIGURE 8.5

Regression analysis for narrowband phase trend of minimum-phase parts extracted from the transfer function used in Fig. 8.4 from Figs. 5 and 6 [12].

The left panel illustrates examples of phase responses for the minimum-phase components where the center frequency is 400 (Hz) with the frequency band of 200 (Hz) as a function of the distance from the source. The distance is normalized to the distance from $r_0 = 0.01$ (m). The phase trend can be seen as the distance increases. The right panel displays the phase trend as a function of r where the phase trend is expressed as the slope $-\partial\Phi/\partial k$ (m). The graph gives the range of data as the center frequency changes by 300 – 900 (Hz) every 100 (Hz) with the frequency band of 200 (Hz). The estimated slopes mostly follow the distance r, as expected to the propagation phase.

Interestingly, the length (or distance from the source in which the propagation phase is preserved) can be estimated based on the wave theory analysis for the reverberant space [3][8]. The length that is called the coherent length is given by

$$r_c \cong \sqrt{\frac{A}{64}}, \qquad \text{(m)} \tag{8.13}$$

where

$$A = \alpha S \cong 0.163\frac{V}{T_R}, \qquad (\text{m}^2) \tag{8.14}$$

where V (m^3) is the room volume, T_R (s) gives the reverberation time (expressed by Eq. (6.17)), and S (m^2) shows the surface of the room. Substituting $V = 183$ (m^3) and $T_R = 1$ (s), $r_c \cong 0.68$ (m) is obtained. Fig. 8.5 confirms the coherent length in which the propagation phase is preserved.

The number of zeros at +infinity for the minimum-phase component in a reverberant response is equal to the number of remaining poles so that the trend of propagation phase might be canceled. The remaining poles characterize the frequency responses, even in the coherent region as a function of the distance from the source and even if the phase trend is lost. This implies that the magnitude frequency response can be a key to perception of distance from the source as far as the propagation phase is not the pure delay. The waveforms of responses may change due to the distance under the minimum-phase criterion even after discarding the initial delay portions that correspond to the propagation phase delay.

8.2 Perception of distance from source in coherent region

8.2.1 Frequency characteristics of transfer function as function of distance in coherent region

An interesting question would be whether the distance from the source can be perceptually estimated, in particular in the coherent region for reverberant space. Determining the distance would be unlikely by listening to sounds in a free field because the loudness of the source signal does not seem to be a key factor.

A candidate of the distance-dependent function might be D_{30} in reverberant space [18]. As shown in Fig. 7.8, however, D_{30} is almost constant in the coherent region. Thus, D_{30} is not able to explain perception of the sound source distance.

Fig. 8.6 displays the magnitude frequency characteristics that depends on the sound source distance and their standard deviations on a dB scale where the frequency range 100 – 8000 (Hz) is taken [19][20]. The normalization at $r = r_0 = 0.01$ (m) in Fig. 8.6 discards the effects of the sound source (loudspeaker) characteristics used for the measurements of the responses. The standard deviation increases uniformly as the distance increases in the coherent region. In the reverberant field far from the sound source the standard deviation is theoretically estimated at around 5.5 (dB) [3][21][22][23]. If the sound source distance can be subjectively determined, then the standard deviation of the magnitude responses might be a possible candidate

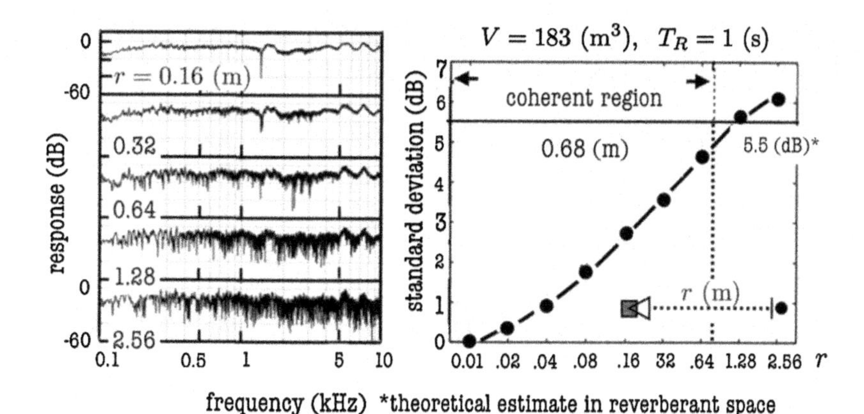

FIGURE 8.6

Magnitude frequency responses of receiving sound (left) and their standard deviations on dB scale (right) as function of distance from the source for a frequency range of $100 - 8000$ (Hz) and normalized at $r = 0.01$ (m) to be 0 (dB).

that follows the perceptual result. The deviations in the magnitude responses can be observed for the minimum-phase part of the transfer function. It implies that the perception of distance would be possible even for the minimum-phase transfer function. On the other hand, the distance perception would be unlikely when only the all-pass component of the transfer function is taken, even in the coherent region.

8.2.2 Subjective evaluation of sound source distance

Subjective tests are reported in [20] on the perception of the sound source distance in the coherent region under diotic listening to speech materials. The diotic listening means binaural listening to a pair of identical signals through headphones in which no binaural information is heard. Fig. 8.7 illustrates the schematic of making speech samples used for the listening tests [20]. Sentences spoken in Spanish and Japanese are recorded by a bilingual female speaker in an anechoic room. The Spanish speech samples without reverberation are convolved with the original (discarding the initial delay), minimum-phase, and all-pass impulse response records at every sound source distance in a reverberation room. The original impulse response is decomposed into the minimum-phase and all-pass components at every receiving position (nine locations). All the synthesized speech materials have a duration of 1 (s) with $1/3$ (s) fade in and out by Hanning windows. The signal energy of all the samples is normalized to unity. Every sample is composed of superposed Spanish and Japanese, which were recorded at different sound source distances. The number of combinations for the pairs of spoken sentences (Spanish and Japanese) is $9 \times 9 = 81$ including the same distance for the three types (original, minimum-phase, and all-pass) of the transfer functions. Six subjects performed the listening tests; they were asked to decide which speech (or language) seemed farther away for every sample under the diotic listening

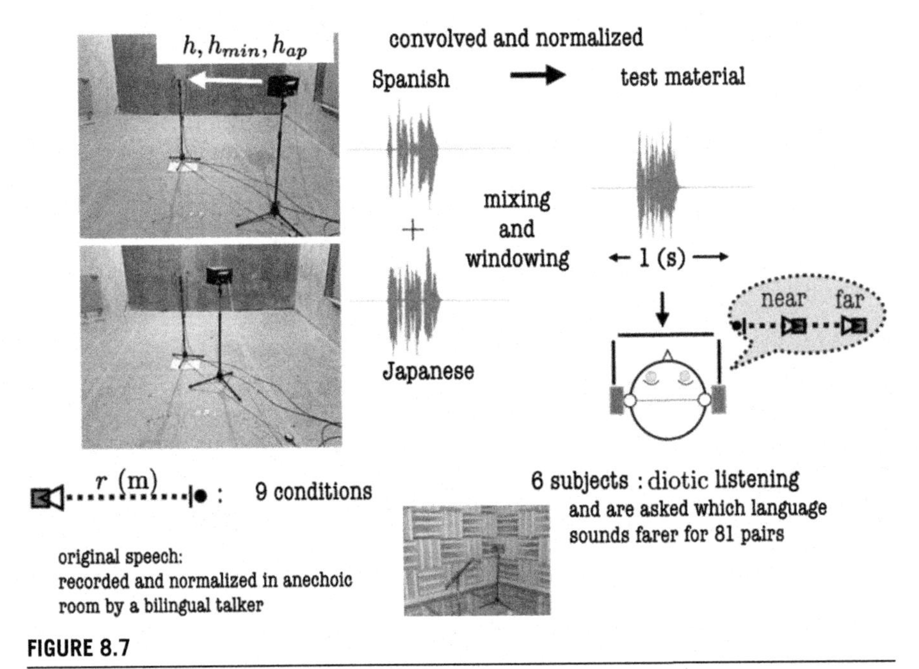

FIGURE 8.7

Experimental conditions for perception of sound source distance in reverberant space from Fig. 4 [20].

through headphones. The subjects could play the test samples as many times as they wanted.

The experimental results [20] are shown in Fig. 8.8. The horizontal axis represents the physical scale, and the vertical axis gives the psychological scale obtained by the experiments. Distance from the source might be perceptually distinguished even in the coherent region under reverberant conditions as far as seeing the results. The impulse response or its minimum-phase component may convey some physical cue to perception of the sound source distance in the coherent region. In contrast, the all-pass component may not exhibit a good correspondence to the sound source distance, as seen in the left panel. The results imply that the standard deviation of the magnitude frequency response might be a possible candidate of a perceptual cue to the distance, as shown in the right panel of the figure. The standard deviation on a dB scale increases uniformly as the sound source distance increases, as presented in Fig. 8.6.

The sound source distance in the coherent region in a reverberant space may be perceptible mostly due to the minimum-phase characteristics of the transfer function or the magnitude spectral responses in the frequency domain. Perception of the source distance outside the coherent region might not be described according to the magnitude spectral behavior such as the standard deviation because the standard deviation can be almost constant independent of the distance from the source [22]. Although masking effects may be included in the tests using samples of two different super-

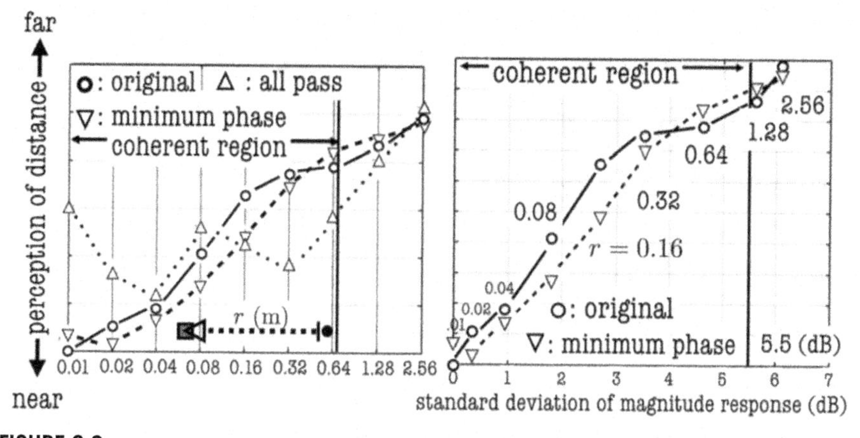

FIGURE 8.8

Perception of sound source distance as a function of distance from the sound source (left panel) or the standard deviation of magnitude response at the receiving position in reverberant space (right panel) from Figs. 5 and 6 [20].

posed languages, superposed sounds are quite likely in recording and reproduction conditions for musical instruments at different distances.

8.2.3 Magnitude frequency characteristics for coherent and reverberation fields

Fig. 8.9 shows the magnitude frequency responses as a function of the distance between source and receiving positions in the reverberant and anechoic rooms where a small loudspeaker system and a microphone are located [19]. Comparing the frequency characteristics in a reverberation room with those for a free field (such as an anechoic room), the effects of reverberation can be clearly seen. However, the difference between the two types of spaces is not as prominent inside the coherent region. In particular, if the responses at $r = 0.01$ are taken (shown in the left and center columns as reference), no significant difference can be seen between the responses in a reverberant space and those in an anechoic room because the responses are mostly due to the direct sound from the loudspeaker. Actually, the frequency responses observed in the anechoic room indicate the frequency characteristics of the loudspeaker itself, and thus the responses mostly trace the spectral envelopes of the frequency characteristics in the reverberant space as mentioned in Fig. 7.6. Source spectral signature can be preserved in the spectral envelope even in a reverberant space although the fine structure of signal components may be deformed by the reverberation.

The spectral change due to the initial echoes (in particular, within 30 (ms)), however, are interesting, as are those for reverberation [19]. The frequency characteristics depend on the sound source distance even in the coherent region because of the initial

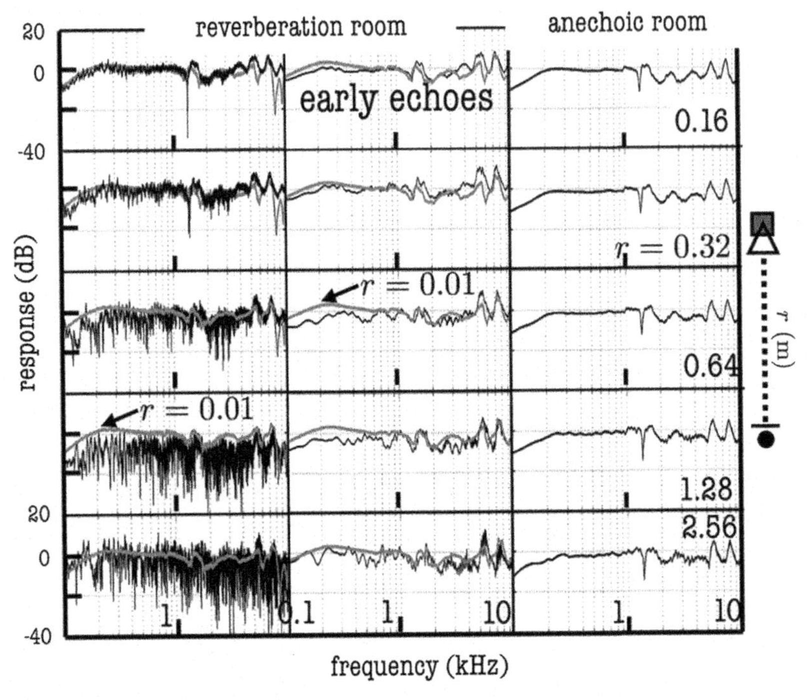

FIGURE 8.9

Magnitude frequency responses for reverberant space (left), initial reflections within 30 (ms) in the reverberant space (center), and for an anechoic room (right) as a function of sound source distance from Figs. 4.11 and 4.12 [6].

echoes. The spatial impression of sound can be sensitive to the initial echoes in the reverberant space [24].

8.3 Auto-correlation analysis of musical sound

8.3.1 Frame-wise auto-correlation functions for musical tones from a piano

Auto-correlation functions exhibit the harmonic structure of musical tones. However, the sound radiated by a sound source is deformed by the reflections and reverberation of the surroundings. Suppose that the power spectrum of a source is $P_S(\omega)$ and the power spectral property of the transfer function from the source to the receiving point $P_H(\omega)$, then the power spectrum of the received sound $P_Y(\omega)$ can be formulated as

$$P_Y(\omega) = P_S(\omega) \cdot P_H(\omega). \tag{8.15}$$

Thus, the auto-correlation function for the received sound can be given by

$$r_y(\tau) = r_s * r_h(\tau), \tag{8.16}$$

where r_s and r_h denote the auto-correlation function of the source and the transfer function.

Fig. 8.10 illustrates samples of the frame-wise auto-correlation functions of a single tone from a piano with a frame length of 30 (ms), where the recording microphone is located in front of an upright piano in a small room.

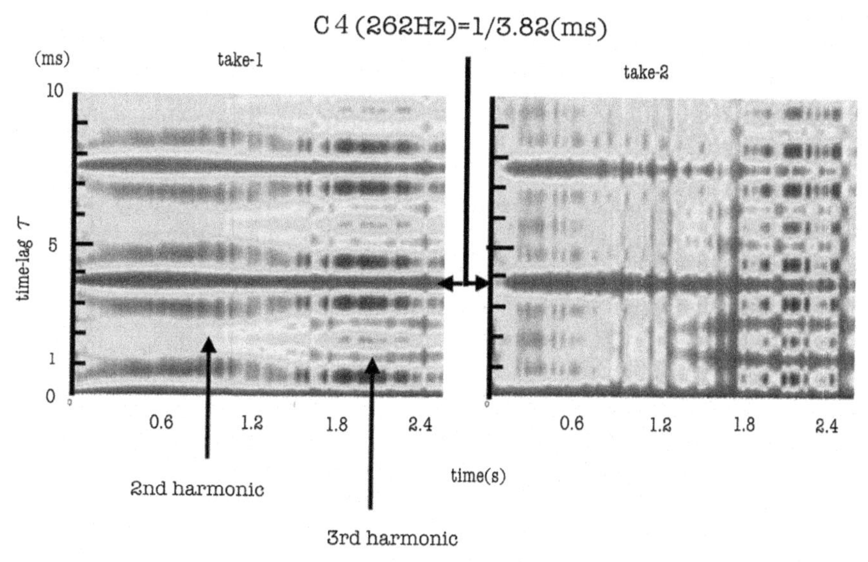

FIGURE 8.10

Samples of frame-wise auto-correlation functions for single tone (C4) from a piano.

Periodic structures are seen in both panels. The period of a single note C4, $1/262 \cong 3.8$ (ms) is read out from the periodical property of the auto-correlation functions. Interestingly, effects of the higher harmonics of the tone on the frame-wise auto-correlation functions can be seen a little after the onset, and the difference between the samples (takes 1 and 2) can be observed in the harmonics [25].

The build-up of the third harmonics on the auto-correlation function recalls the build-up of a compound signal of superposed fundamental, second, and third harmonics. Fig. 8.11 is a schematic of the auto-correlation functions for compound harmonic signals. Frame-wise auto-correlation functions give an estimate of the period and of the periodic structure of a compound signal in the time domain. The functions can show a build-up of higher harmonics as small peaks between the fundamental peaks.

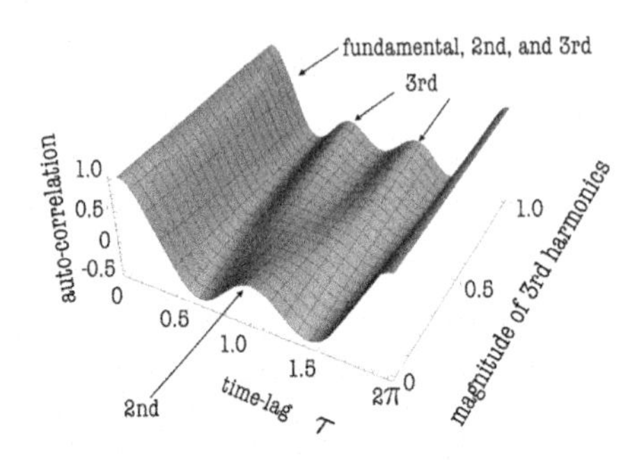

FIGURE 8.11

Build-up of third harmonics in the auto-correlation function of a compound harmonic signal.

8.3.2 Auto-correlation analysis of musical sound in coherent region

Auto-correlation functions express the effects of environments on signal signatures, even in the coherent region, according to Eq. (8.16). Fig. 8.12 shows the auto-correlation functions from $\tau = 0$ to 20 (ms) for a brief musical excerpt (30 (ms)) taken from G. F. Handel's "Water Music Suite" [26] reproduced in the reverberant space (left) and in the anechoic room (right) [19][20][27].

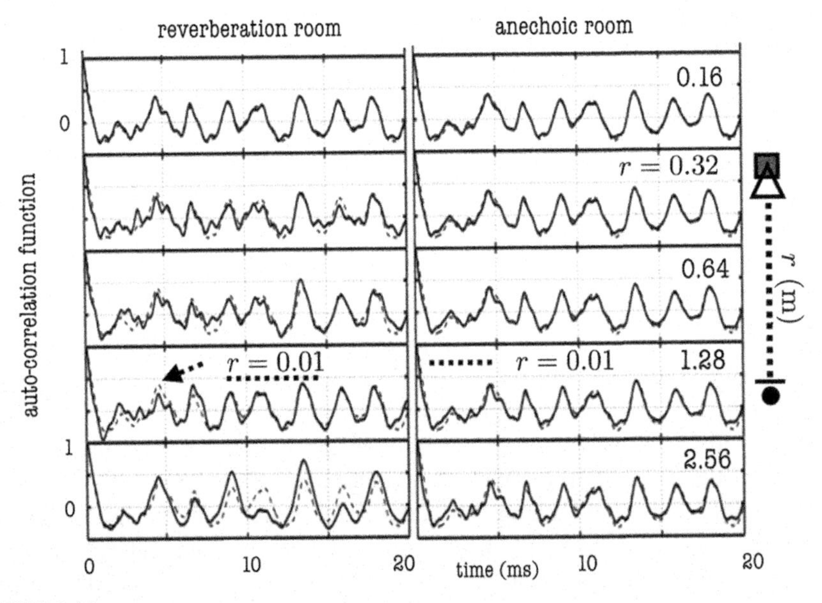

FIGURE 8.12

Initial portion of auto-correlation functions for a short excerpt of music (30 (ms)) in reverberant space (left) and in an anechoic room (right).

Taking the auto-correlation function at $r = 0.01$ (m) as a reference shown by the broken lines in each panel, then the effects of the sound source distance on the auto-correlation functions are more clearly seen in the reverberant space than in the anechoic room. Fig. 8.13 summarizes the total squared sum of the differences in the auto-correlation functions within 20 (ms) of the reference in each panel.

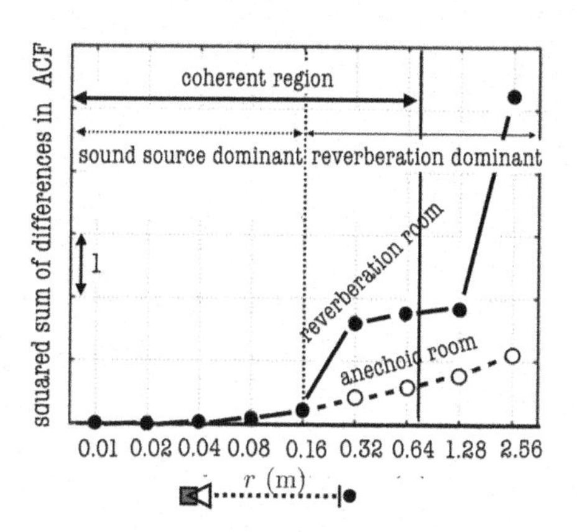

FIGURE 8.13

Squared sum of differences in auto-correlation functions in a reverberant space or an anechoic room.

The difference due to the sound source distance in the anechoic room represents the effects of the loudspeaker system used in this recording. On the other hand, the variations for the coherent region in the reverberant space are, more or less, similar to those in the anechoic room; however, the deformation of the auto-correlation functions is prominent outside the coherent region. In the example shown in Fig. 8.13, $r = 0.16$ (m) seems a separation of the effects of the source and environment. A subjective evaluation of the separation is an interesting issue [19].

8.4 Subjective evaluation of musical sound in coherent region

8.4.1 Conditions of subjective evaluation

Auto-correlation functions exhibit spectral deformation of source signals due to the sound source distance even in the coherent region. A subjective evaluation of the deformation by listening would be interesting. Loudness is a subjective quality of musical sounds. A study of subjective evaluations of the coherent region is reported in the references [19][20][27]. Fig. 8.14 shows the experimental conditions for a subjective evaluation of loudness. A short passage of music from Handel, the Water Music Suit is taken for the tests [26]. The signal level of each test material is normalized, and fading in and out with Hanning windows is applied for the onset and

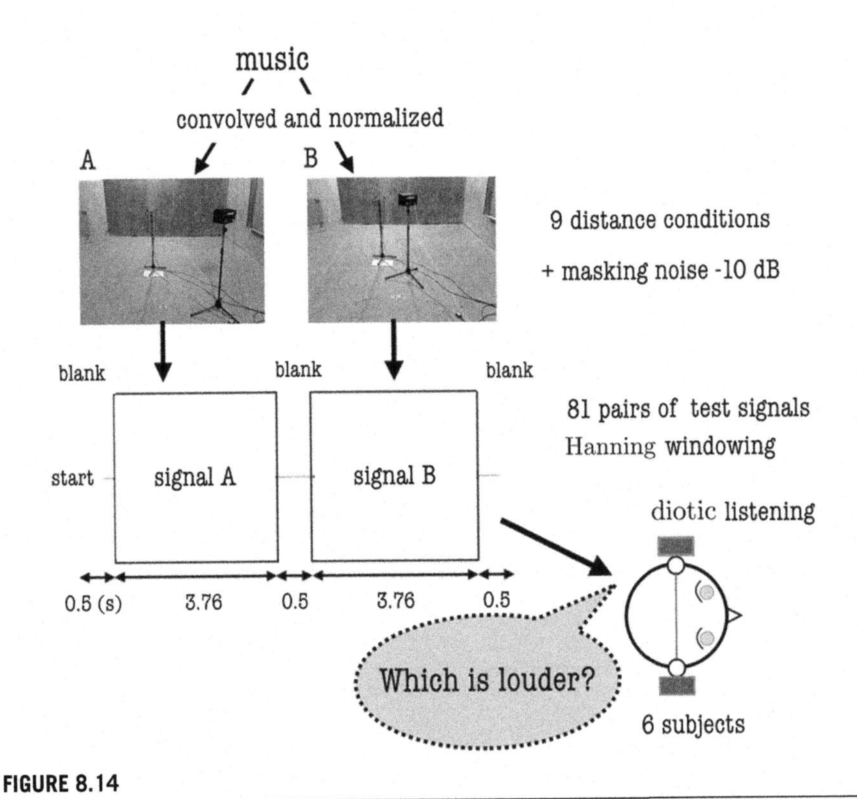

FIGURE 8.14

Experimental conditions for subjective evaluation of loudness by paired comparison test.

offset portions, respectively. All the signals are band-limited to the frequency range from 100 (Hz) to 10 (kHz). A random noise that is made by convolution of a white noise and the impulse response of the reverberant space is added to the test signals. It would make a subject's decision stable by masking the effects of the long reverberation. Six subjects who performed the experiments listened to the pairs of test signals through headphones under diotic listening conditions. They were asked which of the two signals sounded louder. The subjects could play the signal pairs as many times as they wanted by preferred volume.

8.4.2 Subjective evaluation of loudness for musical sounds

Figs. 8.15 presents subjective evaluation scores for loudness by the paired comparison tests. The horizontal axis represents the distance from the source and the vertical axis shows the psychological scale assuming Thurston's case V in the historical references [28][29]. The bars in the figure suggest the 95% confidence intervals according to the reference [30]. The left panel presents the results under reverberant conditions, while the right panel gives the case in an anechoic room as a reference.

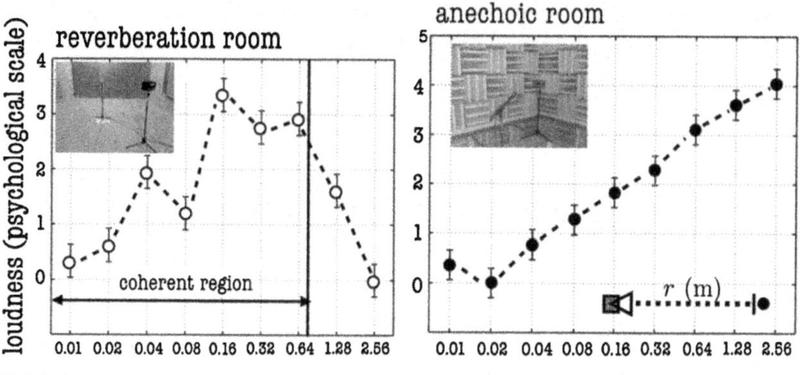

FIGURE 8.15

Subjective scores for loudness of musical sounds under reverberant (left) and anechoic (right) conditions from Fig. 3.6 [19].

A noticeable difference can be seen outside the coherent region between the left and right panels in Fig. 8.15. Loudness is at a maximum at $r = 0.16$ (m) under reverberant conditions, while loudness increases almost monotonically in the anechoic room as the sound source distance inccreases. The increase in loudness in the anechoic room implies effects due to the frequency characteristics of the loudspeaker system used in the experiments, which depends on the distance from the source. Similarly, the increase for $r \leq 0.16$ in the coherent region under the reverberant condition can be understood as the source effect.

The separation between the increase and the decrease in the loudness at $r = 0.16$ (m) in the example of Fig. 8.15 may show the boundary on the sound source distance under reverberation conditions. The perceptual effects of reflections on the loudness can be seen in the frequency domain in the coherent region. In contrast, the effects can be prominent in the time domain outside the coherent region instead of the frequency domain. The decrease in the loudness reveals that reverberation may not be helpful in the enhancement of loudness outside the coherent region. Recalling that D_{30} is mostly preserved in the coherent region, interestingly, Fig. 8.15 partly confirms the concept of D_{30} [31][32].

8.5 Envelope analysis and intelligibility of speech in the coherent region

8.5.1 Listening tests for speech samples recorded in the coherent region

Speech intelligibility depends on the envelope of the waveform, as described in Chapter 7. In the coherent region the intelligibility would be almost perfect because the direct sound is dominant even in reverberant space. However, daily experience implies that the quality of speech changes with the distance from the source, even in the coherent region. The subjective evaluation of speech intelligibility is reported [19][33] by paired comparison tests. Fig. 8.16 presents the experimental conditions for the comparison tests on speech intelligibility, similar to the conditions in Fig. 8.14 for the loudness of musical sound.

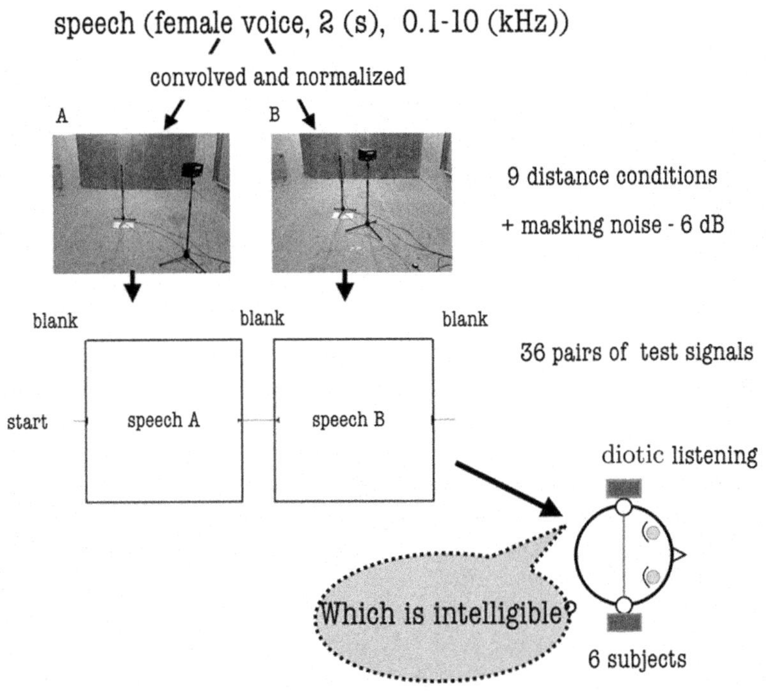

FIGURE 8.16

Experimental conditions for paired comparison tests for speech intelligibility in the coherent region from Fig. 12.17 [7].

Six subjects are asked which of two signals is more intelligible. The random noise convolved with the impulse response in the reverberant room is added to sound pairs so that the effect of long echoes might be avoided, and thus the comparison would be easier for the subjects.

Fig. 8.17 illustrates the results of the paired comparison tests. The bars in the figure suggest the 95% confidence intervals according to the reference [30]. The speech is more intelligible as the distance increases as long as the distance is within $r \leq 0.16$ in the coherent range under reverberant conditions; however, speech samples rapidly become unintelligible as the distance exceeds the coherent range. In contrast, the speech quality can be enhanced in an anechoic room as the distance increases. Again, the difference in the condition $r = 0.16$ (m) in the reverberant room is interesting. The loss of intelligibility in the reverberant space seems to correspond to the enve-

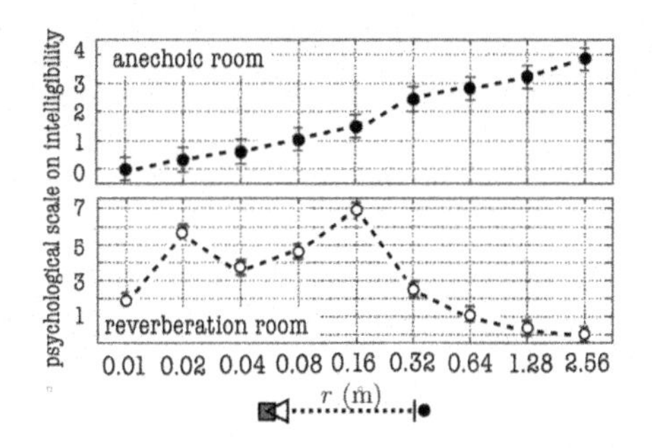

FIGURE 8.17

Results of paired comparison tests on speech intelligibility under reverberant conditions (lower panel) and in an anechoic room as a reference (upper panel) from Fig. 1 [33].

lope energy that decreases as the distance exceeds the coherent range, as shown in Fig. 7.9.

8.5.2 Power spectral behavior of squared envelope of reverberant speech

A power spectral analysis of squared narrowband envelopes was presented in Fig. 7.9 in a reverberant space. Fig. 8.18 illustrates the prominent spectral components of narrowband speech envelopes similar to Fig. 7.9 including the results for the anechoic room as a reference.

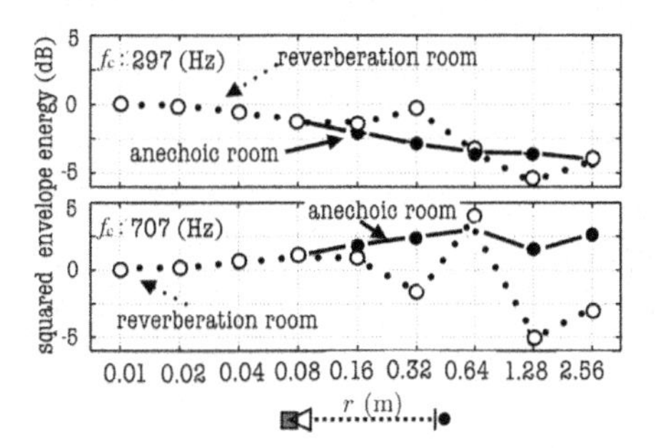

FIGURE 8.18

Energy of prominent envelopes of speech under reverberant conditions or in an anechoic room as function of distance from the source from Fig. 4 [33].

Comparing the results between the reverberant and anechoic rooms, the dependence on the sound source distance seems to correspond to the envelope energy at 707 (Hz) as a function of the distance. The monotonic increase in the envelope energy for the anechoic room implies that the loudspeaker system may enhance the intelligibility of speech. On the other hand, the enhancement for reverberant speech due to the loudspeaker system and early reflections might be expected only in the region $r \leq$

0.16 (m) in the coherent region as shown by the example in Fig. 8.18. The spectral characteristics of the envelopes are largely deformed by the reflections outside the coherent region.

Narrowband envelopes of music samples are enhanced or deformed under reverberant conditions, as are the speech samples. Fig. 8.19 displays the frequency dependence of sub-band envelope energy, as a function of the sound source distance, for the music sample used for the loudness test displayed in Fig. 8.15.

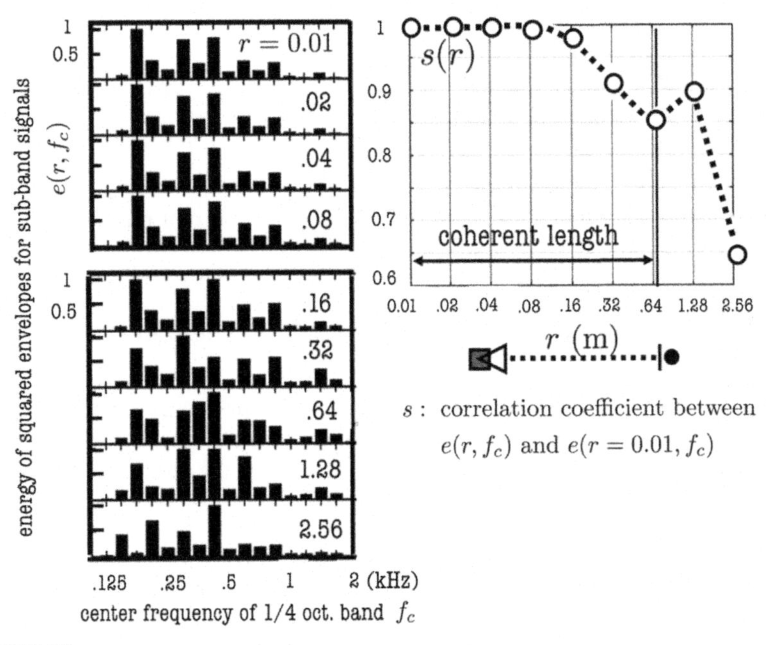

FIGURE 8.19

Similar to Fig. 8.18, but for the music sample used in Fig. 8.15 from Figs. 11 and 12 [27].

In addition to the auto-correlation analysis in Fig. 8.13, the energetic view of sub-band envelopes exhibits the deviations from the reference at $r = 0.01$. Envelopes can be enhanced mostly depending on the characteristics of the loudspeaker system preserving the signals in the envelopes when the sound source distance is within 0.16 (m). The right panel shows the similarity of the frequency dependence of the energy of sub-band envelopes by taking cross-correlation coefficients between the observed envelopes at r and $r = 0.01$ (the reference). The similarity can be mostly preserved when $r < 0.16$; however, it breaks when r exceeds 0.16. The similarity makes the subjective scores higher for loudness or intelligibility, while the dissimilarity decreases the scores.

Enhancement of preference scores are reported in references [19][20][27][34]. Intensive study has been performed in reference [24] on theoretical and physiological research including the preference of sound fields, taking the example of concert hall acoustics. The conditions taken in this chapter may be closely related to those for the referenced work; however, the approaches stated in the examples here are applicable to sound recording engineering.

8.6 Exercises

1. Confirm Eqs. (8.10) and (8.14).

2. Deriving Eq. (8.13) is a little complicated. If some readers are interested in the process, try to trace the steps according to references [3][8].

3. The theoretical estimation of the standard deviation of magnitude frequency characteristics of sound pressure responses in a reverberant space is around 5.5 (dB) on the decibel scale. If some of the readers are interested in the theoretical estimation, then follow the process according to references [3][22][23].

4. Confirm Eqs. (8.15) and (8.16). Where is the effect of the phase spectral components?

5. Show an example in which the period (or fundamental frequency) of a compound signal can be estimated by the auto-correlation functions, even if the fundamental or some of harmonics are lost (see sections 1.1.3 and 1.1.4).

6. Explain the terminologies listed below:
(1) propagation phase (2) linear phase (3) pure delay
(4) accumulated phase (5) reverberation phase (6) coherent region
(7) sign of residue (8) pole-line (9) minimum-phase zero
(10) nonminimum-phase zero (11) symmetric pair of zeros with respect to the pole-line (12) distance between the pole-line and real frequency axis
(13) damping factor (14) standard deviation of magnitude frequency response
(15) diotic listening
(16) spectral envelope (17) auto-correlation function (18) tangential wave
(19) cross-correlation coefficient

References

[1] J. Blauert, N. Xiang, Acoustics for Engineers, Springer, 2008.
[2] M. Tohyama, T. Koike, Fundamentals of Acoustic Signal Processing, Academic Press, 1998.
[3] M. Tohyama, Sound and Signals, Springer, 2011.
[4] R.H. Lyon, Progressive phase trends in multi-degree-of-freedom systems, J. Acoust. Soc. Am. 73 (4) (1983) 1223–1228.
[5] R.H. Lyon, Range and frequency dependence of transfer function phase, J. Acoust. Soc. Am. 76 (5) (1984) 1435–1437.

[6] M. Tohyama, Waveform Analysis of Sound, Springer, 2015.
[7] M. Tohyama, Sound in the Time Domain, Springer, 2017.
[8] P.M. Morse, R.H. Bolt, Sound waves in rooms, Rev. Mod. Phys. 16 (1944) 69–150.
[9] M. Tohyama, R.H. Lyon, Zeros of a transfer function in a multi-degree-of-freedom system, J. Acoust. Soc. Am. 86 (5) (1989) 1854–1863.
[10] R. Nelson, Probability, Stochastic Processes, and Queueing Theory, Springer, 1995.
[11] R.H. Lyon, Statistical analysis of power injection and responses in structures and rooms, J. Acoust. Soc. Am. 45 (3) (1969) 545–565.
[12] Y. Takahashi, M. Tohyama, Y. Yamasaki, Phase response of transfer functions and coherent field in a reverberation room, Electron. Commun. Jpn., Part 3 90 (4) (2007) 1–8.
[13] M. Tohyama, R.H. Lyon, T. Koike, Reverberant phase in a room and zeros in the complex frequency plane, J. Acoust. Soc. Am. 89 (4) (1991) 1701–1707.
[14] M. Tohyama, R.H. Lyon, T. Koike, Phase variabilities and zeros in reverberant transfer function, J. Acoust. Soc. Am. 95 (1) (1994) 286–296.
[15] M. Tohyama, H. Suzuki, Y. Ando, The Nature and Technology of Acoustic Space, Academic Press, London, 1995.
[16] M. Tohyama, Response statistics in rooms, in: M.J. Crocker (Ed.), Encyclopedia of Acoustics 2(77), John-Wiley and Sons, Inc., Chester, 1997, pp. 913–923.
[17] M. Tohyama, Room transfer function, in: Handbook of Signal Processing in Acoustics; Part 3, Engineering Acoustics 75, 2008, pp. 1381–1402.
[18] P. Zaholic, Direct-to-reverberant energy ratio sensitivity, J. Acoust. Soc. Am. 112 (5) (2002) 2110–2117.
[19] Y. Hara, Sound perception and temporal-spectral characteristics in sound field near sound source, PhD Thesis No. 126, Kogakuin University, Tokyo, Japan, Jan. 27, 2014.
[20] Y. Hara, Y. Takahashi, H. Nomura, M. Tohyama, K. Miyoshi, Perception of sound source distance and loudness in a coherent field, in: Acoustics'08, 2008, p. 2196, 4pPPd6.
[21] M.R. Schroeder, Die attistischen Parameter des Frequenzkurven von grossen Raeumen, Acustica (Beiheft 2) 4 (1954) 594–600.
[22] M.R. Schroeder, Statistical parameters of the frequency response curves in large rooms, J. Audio Eng. Soc. 35 (5) (1987) 299–306.
[23] K.J. Ebeling, Properties of random wave fields, in: Physical Acoustics XVII, Academic Press, London, 1984, pp. 233–310.
[24] Y. Ando, Signal Processing in Auditory Neuroscience, Temporal and Spatial Features of Sound and Speech, Elsevier, 2018.
[25] S. Ohno, Russian Pianism (in Japanese), Yamaha Music. Entertainment Holdings, 2019.
[26] Denon Professional Test CD, Track 9 II, Anechoically recorded samples for evaluation: No. 6-Water Music Suite, Bars 1-11, and Track 12 II, Anechoically recorded samples for evaluation: Symphony No. 4 in E minor, Op. 98, Bars 354-362.
[27] Y. Hara, H. Nomura, M. Tohyama, K. Miyoshi, Subjective evaluation for music recording positions in a coherent region of a reverberant field, in: Audio Engineering Society 124th Convention 7446, Amsterdam, 2008.
[28] T.T. Thurstone, Law of comparative judgement, Psychol. Rev. 34 (4) (1927) 273–286.
[29] T.T. Thurstone, Psychophysical analysis, Am. J. Psychol. 38 (1927) 368–389.
[30] E.D. Montag, Empirical formula for creating error bars for the method of paired comparison, J. Electron. Imaging 15 (1) (2006) 010502.
[31] R. Thiele, Richtungsverteilung und Zeitfolge der Schallrueckwuerfe in Raeumen, Acustica 3 (1953) 291–302.
[32] T.J. Schlutz, Acoustics of the concert hall, IEEE Spectr. (June 1965) 56–67.
[33] Y. Hara, Y. Takahashi, K. Miyoshi, Narrow-band spectral energy analysis close to sound source, J. Inst. Electron. Inf. Comm. Eng. Jpn. J 97-A (3) (2014) 221–223.
[34] Y. Hara, Y. Takahashi, H. Nomura, M. Tohyama, K. Miyoshi, Preference of the transfer functions for music recording in as coherent region of a reverberant field, in: CAS-01-007, 19th Int. Cong. Acoust., Madrid, ICA, 2007.

CHAPTER

Spatial impression and binaural sound field

CONTENTS

Acoustic Signals and Hearing. https://doi.org/10.1016/B978-0-12-816391-7.00017-6

9.1 Sound image localization and binaural condition of sound

9.1.1 Head-related transfer functions and linear equations for sound image rendering

Suppose a listener and a point source as shown in Fig. 9.1.

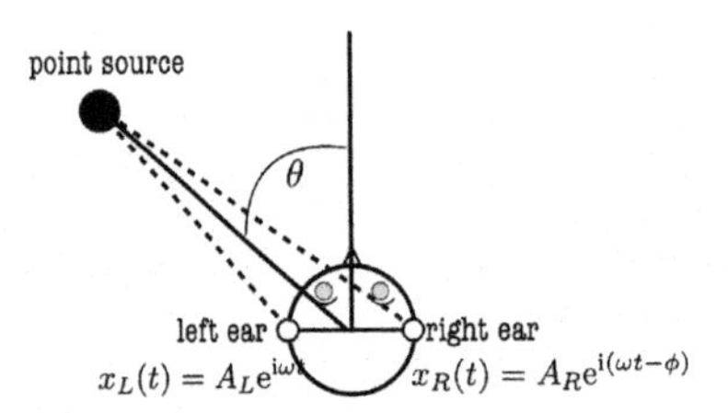

FIGURE 9.1

Location of listener and sound source.

Assuming the left-ear signal to be $x_L(t) = A_L e^{i\omega t}$, then the right-ear signal can be expressed as $x_R(t) = A_R(\omega)e^{i(\omega t-\phi(\omega))}$. The ratio $|A_L/A_R|$ and difference $\phi(\omega)$ are called the binaural magnitude ratio and phase difference, respectively. The binaural conditions such as the magnitude and phase differences are significant to render the sound image localized at the source position. The transfer function from a sound source to the right or left ear entrance (or ear drum) is customarily called the head-related transfer function (HRTF) [1][2][3][4]. The HRTF is denoted by $H(z^{-1})$ in the $z-$transform of the corresponding impulse response $h(n)$.

Suppose a pair of loudspeakers and a listener as shown in Fig. 9.2.

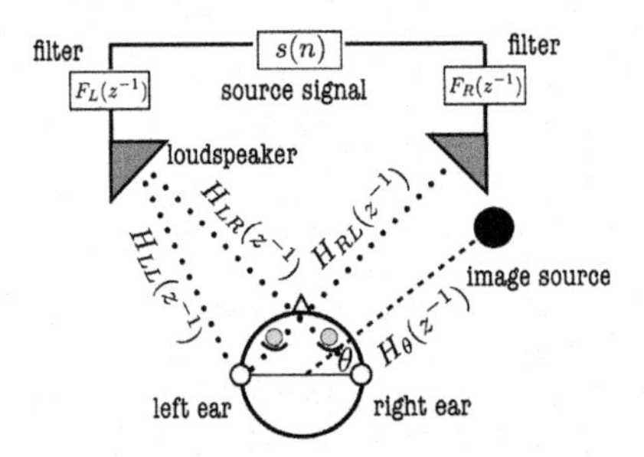

FIGURE 9.2

Filters and loudspeaker arrangement for binaural reproduction.

Assuming the HRTFs from the loudspeakers to the ear entrances of the listener such as $H_{LR}\left(z^{-1}\right)$ from the left loudspeaker to the right ear, the $z-$transforms of a binaural pair of signals are written as

$$S_L\left(z^{-1}\right) = S\left(z^{-1}\right)\left(H_{LL}\left(z^{-1}\right) + H_{RL}\left(z^{-1}\right)\right) \tag{9.1}$$

$$S_R\left(z^{-1}\right)=S\left(z^{-1}\right)\left(H_{LR}\left(z^{-1}\right)+H_{RR}\left(z^{-1}\right)\right),$$

where $S\left(z^{-1}\right)$ denotes the z−transform of the signal $s(n)$. Setting a pair of binaural filters $F_L\left(z^{-1}\right)$ and $F_R\left(z^{-1}\right)$, as shown in Fig. 9.2, the binaural pair of receiving signals is controllable by using the filters. The receiving signals through the binaural filters can be formulated as

$$\hat{S}_L\left(z^{-1}\right)=S\left(z^{-1}\right)\left[F_L\left(z^{-1}\right)H_{LL}\left(z^{-1}\right)+F_R\left(z^{-1}\right)H_{RL}\left(z^{-1}\right)\right] \tag{9.2}$$
$$\hat{S}_R\left(z^{-1}\right)=S\left(z^{-1}\right)\left[F_L\left(z^{-1}\right)H_{LR}\left(z^{-1}\right)+F_R\left(z^{-1}\right)H_{RR}\left(z^{-1}\right)\right].$$

Assume a binaural pair of the HRTFs $H_{\theta L}$ and $H_{\theta R}$ for a sound source. Setting the binaural pair of filters $F_{\theta L}$ and $F_{\theta R}$ that satisfy the linear equations

$$H_{\theta_L}\left(z^{-1}\right)=F_{\theta L}\left(z^{-1}\right)H_{LL}\left(z^{-1}\right)+F_{\theta R}\left(z^{-1}\right)H_{RL}\left(z^{-1}\right) \tag{9.3}$$
$$H_{\theta_R}\left(z^{-1}\right)=F_{\theta L}\left(z^{-1}\right)H_{LR}\left(z^{-1}\right)+F_{\theta R}\left(z^{-1}\right)H_{RR}\left(z^{-1}\right)$$

under the given HRTFs (H_{LL}, H_{RL}, H_{LR}, and H_{RR}), then the listener might perceive the sound image along the θ−direction, as shown in Fig. 9.2. The pair of filters is given by [5]

$$F_{\theta_L}\left(z^{-1}\right)=\frac{H_{\theta L}\left(z^{-1}\right)H_{RR}\left(z^{-1}\right)-H_{\theta R}\left(z^{-1}\right)H_{RL}\left(z^{-1}\right)}{H_{RR}\left(z^{-1}\right)H_{LL}\left(z^{-1}\right)-H_{RL}\left(z^{-1}\right)H_{LR}\left(z^{-1}\right)} \tag{9.4}$$
$$F_{\theta_R}\left(z^{-1}\right)=\frac{H_{\theta R}\left(z^{-1}\right)H_{LL}\left(z^{-1}\right)-H_{\theta L}\left(z^{-1}\right)H_{LR}\left(z^{-1}\right)}{H_{RR}\left(z^{-1}\right)H_{LL}\left(z^{-1}\right)-H_{RL}\left(z^{-1}\right)H_{LR}\left(z^{-1}\right)}.$$

If a pair H_{θ_L} and H_{θ_R} (which do not correspond to any direction θ) is taken, it would be intriguing to know what sound image is presented to the listener. Differences between a spatially localized sound image by loudspeakers and a laterally shifted in-head image through headphones might show an example of the cases.

9.1.2 Binaural filters for magnitude and time differences between two receiving points

Suppose a conceptual model for a two-channel reproduced field, as shown in the left panel of Fig. 9.3. The impulse response from the virtual source to the receiving positions can be rewritten as shown in the right panel

$$g_L(n)=b\cdot\delta(n-K-J)\ \text{and}\ g_R(n)=\delta(n-K), \tag{9.5}$$

where b $(0<b<1)$ denotes the magnitude difference in the receiving pair for the virtual source. The z-transforms for the impulse responses given by Eq. (9.5) become

$$G_L\left(z^{-1}\right)=b\cdot z^{-K}z^{-J}\ \text{and}\ G_R\left(z^{-1}\right)=z^{-K}. \tag{9.6}$$

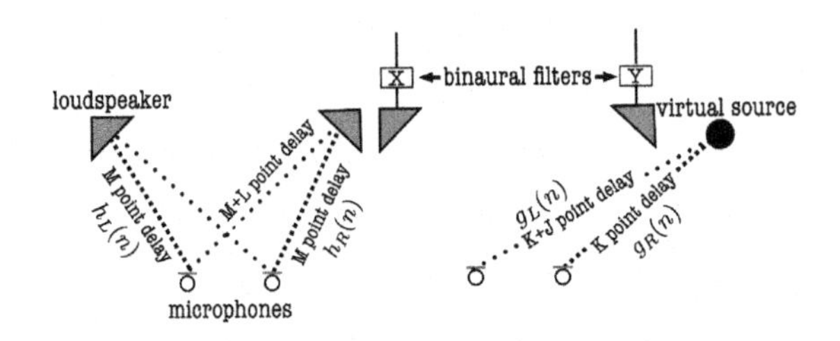

FIGURE 9.3

Simplified sound path and binaural filters for sound image projection.

The equations that the binaural filters $X(z^{-1})$ and $Y(z^{-1})$ satisfy can be formulated as

$$G_L\left(z^{-1}\right) = X\left(z^{-1}\right)z^{-M} + Y\left(z^{-1}\right)a \cdot z^{-(M+L)} = b \cdot z^{-(K+J)} \tag{9.7}$$

and

$$G_R\left(z^{-1}\right) = X\left(z^{-1}\right)a \cdot z^{-(M+L)} + Y\left(z^{-1}\right)z^{-M} = z^{-K}, \tag{9.8}$$

where a $(0 < a < 1)$ gives the magnitude difference in the receiving signal pair from the loudspeaker system shown in the left panel of Fig. 9.3. Solving the equations of $X(z^{-1})$ and $Y(z^{-1})$, then

$$X\left(z^{-1}\right) = \frac{b \cdot z^{-J} - a \cdot z^{-L}}{z^{K-M}\left(1 - a^2 \cdot z^{-2L}\right)} = \frac{N_X(z^{-1})}{D(z^{-1})} \tag{9.9}$$

and

$$Y\left(z^{-1}\right) = \frac{1 - a \cdot b \cdot z^{-(J+L)}}{z^{K-M}\left(1 - a^2 \cdot z^{-2L}\right)} = \frac{N_Y(z^{-1})}{D(z^{-1})} \tag{9.10}$$

are obtained. The binaural filters are causal and stable in the time domain when $K > M$ and $0 < |a| < 1$ for the virtual source behind the loudspeakers.

9.1.3 Cross-talk cancelation and spectral equalization by binaural filters

Binaural filters derived as in Eqs. (9.9) and (9.10) imply a concept different from the conventional mixing technologies according to the inter-aural level differences. Substituting the pair of binaural filters given by Eqs. (9.9) and (9.10) for the simultaneous equations displayed by Eqs. (9.7) and (9.8), then

$$G_L\left(z^{-1}\right) = \frac{z^{-M}}{D(z^{-1})}\left(bz^{-J} - az^{-L} + az^{-L} - a^2bz^{-(J+2L)}\right) = bz^{-K}z^{-J} \tag{9.11}$$

and

$$G_R\left(z^{-1}\right)=\frac{z^{-M}}{D(z^{-1})}\left(abz^{-(J+L)}-a^2z^{-2L}+1-abz^{-(J+L)}\right)=z^{-K} \quad (9.12)$$

are derived, where

$$D(z^{-1})=z^{K-M}\left(1-a^2z^{-2L}\right). \quad (9.13)$$

The binaural filters are determined so that the transfer functions might hold well for the virtual source at the binaural listening points. The two-channel reproduction cancels cross-talk on the two-channel signals at the listening positions by using the pair of binaural filters [6].

The cancelation can be seen in Eqs. (9.11) and (9.12). Spectral distortion such as $1-a^2z^{-2L}$, however, a result of the cancelation, although no effects on the binaural information such as binaural spectral differences are left. The spectral distortion can be equalized by the denominator of the binaural filters, which are common for the pair of binaural filters [5].

The numerators N_X and N_Y defined in Eqs. (9.9) and (9.10) make the binaural differences due to the virtual source. On the other hand, the common denominators do not have binaural information, but control the monaural information. Take only the numerators of the binaural filters N_X and N_Y by discarding the denominator [5]. The binaural responses become

$$\hat{G}_L\left(z^{-1}\right)=G_L\left(z^{-1}\right)D(z^{-1})=bz^{-(K+J)}\left(1-a^2z^{-2L}\right) \quad (9.14)$$

and

$$\hat{G}_R\left(z^{-1}\right)=G_R\left(z^{-1}\right)D(z^{-1})=z^{-K}\left(1-a^2z^{-2L}\right). \quad (9.15)$$

The binaural difference due to the virtual source can be preserved even after discarding the denominator of the binaural filters.

Binaural spectral variation can be equalized by the denominators of the binaural filters. Take a minimum-phase filter $E(z^{-1})$ as the denominator instead of $D(z^{-1})$. The inverse filter of a minimum-phase filter is realizable as a causal filter in the time domain [5]. The binaural responses can be written as

$$\overline{G}_L\left(z^{-1}\right)=G_L\left(z^{-1}\right)\frac{D(z^{-1})}{E(z^{-1})}=G_L\left(z^{-1}\right)F(z^{-1}) \quad (9.16)$$

$$\overline{G}_R\left(z^{-1}\right)=G_R\left(z^{-1}\right)F(z^{-1}),$$

where the binaural information (or the ratio G_L/G_R) can be preserved once more. The monaural modification can be equalized by selecting the filter $E(z^{-1})$. The pair of equations for $\overline{G}_L\left(z^{-1}\right)$ and $\overline{G}_R\left(z^{-1}\right)$ implies a possibility in controlling perception of sound source distance, recalling Chapter 8.

9.1.4 Gain control of inter-aural magnitude difference

Controlling magnitude differences between a pair of two-channel signals can also be formulated in terms of the binaural filters. The spectral representation for the binaural filters can be written

$$X\left(e^{-i\Omega}\right) = \frac{b \cdot e^{-i\Omega J} - a \cdot e^{-i\Omega L}}{e^{-i\Omega(M-K)}\left(1 - a^2 e^{-i\Omega 2L}\right)}$$
$$Y\left(e^{-i\Omega}\right) = \frac{1 - a \cdot b e^{-i\Omega(J+L)}}{e^{-i\Omega(M-K)}\left(1 - a^2 e^{-i\Omega 2L}\right)} \quad (9.17)$$

by substituting $z = e^{i\Omega}$ for Eqs. (9.9) and (9.10). Taking the magnitude of the filters by setting 0-phase (or discarding the phase component),

$$\tilde{G}_L\left(e^{-i\Omega}\right) = \left|X\left(e^{-i\Omega}\right)\right| e^{-i\Omega M} + \left|Y\left(e^{-i\Omega}\right)\right| a \cdot e^{-i\Omega M} e^{-i\Omega L}$$
$$\tilde{G}_R\left(e^{-i\Omega}\right) = \left|X\left(e^{-i\Omega}\right)\right| a \cdot e^{-i\Omega M} e^{-i\Omega L} + \left|Y\left(e^{-i\Omega}\right)\right| e^{-i\Omega M} \quad (9.18)$$

are obtained according to Eqs. (9.7) and (9.8). An interesting question is whether the ratio b is reserved, even after discarding the phase components from the binaural filters. The binaural difference becomes

$$\frac{\tilde{G}_L\left(e^{-i\Omega}\right)}{\tilde{G}_R\left(e^{-i\Omega}\right)} = \frac{\left|b e^{-i\Omega J} - a e^{-i\Omega L}\right| + a e^{-i\Omega L}\left|1 - a \cdot b e^{-i\Omega(J+L)}\right|}{a e^{-i\Omega L}\left|b e^{-i\Omega J} - a e^{-i\Omega L}\right| + \left|1 - a \cdot b e^{-i\Omega(J+L)}\right|}$$
$$\cong \left|b - a e^{-i\Omega(L-J)}\right| + a e^{-i\Omega L} \cong b - ia\Omega L \cong b \quad (9.19)$$

assuming a low-frequency range, where $a \cdot b << 1,\ b > a$, and

$$e^{-i\Omega(L-J)} \cong 1, \quad e^{-i\Omega L} \cong 1 - i\Omega L. \quad (9.20)$$

According to the assumption $b > a$, the virtual source can be rendered between the pair of loudspeakers, even after removing the phase components from the binaural filters, when the assumptions described above including Eq. (9.20) hold.

On the other hand, b can be lower than a for a virtual source outside the range between the pair of loudspeakers. However, taking the anti-phase pair of the filters such that $-|X|$ and $|Y|$, then the binaural difference becomes

$$\frac{\overline{\tilde{G}}_L\left(e^{-i\Omega}\right)}{\overline{\tilde{G}}_R\left(e^{-i\Omega}\right)} \cong -|b - a| + a(1 - i\Omega L) \cong b, \quad (9.21)$$

where $b < a$. The anti-phase pair of binaural filters might render a virtual source slightly outside the loudspeakers' region.

9.2 Cross-correlation functions between a pair of receiving sounds

9.2.1 Cross-correlation for sound pressure in sinusoidal wave field

Suppose a sound source that radiates a sinusoidal wave into a room. Taking a pair of receiving positions as shown in Fig. 9.4, the sinusoidal wave can be written as

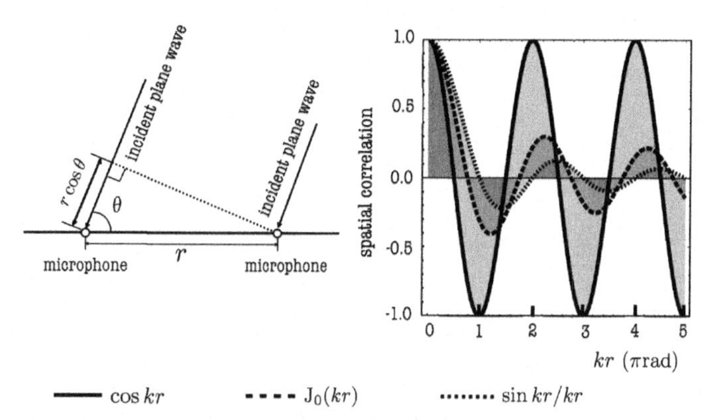

FIGURE 9.4

Pair of microphones in sinusoidal wave field and examples of spatial correlation functions.

$$p_1(t) = A\cos\omega t \text{ and } p_2(t) = A\cos(\omega t - \phi) \tag{9.22}$$

at the receiving positions, respectively, where ω denotes the angular frequency (rad/s) of the sinusoidal wave. Assuming a single sinusoidal wave coming to the receiving points, the phase difference ϕ (rad) is given by

$$\phi = kr\cos\theta \quad \text{(rad)} \tag{9.23}$$

following Fig. 9.4, where r gives the distance (m) between the pair of receiving positions, θ denotes an incident angle for the plane wave coming into the plane on which the receiving positions are located, and k (rad/m) is the wave number of the sinusoidal wave such that ω/c (rad/m), and c (m/s) gives the speed of sound.

Suppose a random sound field in which the angle of the incident wave is equally likely in three-dimensional space. Products of pairs of the receiving signals make an ensemble of random signals,

$$R_s = P_1(t) \cdot P_2(t - \tau), \tag{9.24}$$

where R_s represents the ensemble of the products from Eq. (9.24).

Taking the ensemble average of the products,

$$r_s = \mathrm{E}[R_s] = \frac{A^2}{2}\mathrm{E}[\cos(kr\cos\theta)\,(\cos(2\omega t - \omega\tau) + \cos\omega\tau)] \tag{9.25}$$

$$+\frac{A^2}{2}\mathrm{E}[\sin(kr\cos\theta)\,(\sin(2\omega t-\omega\tau)-\sin\omega\tau)],$$

where $\phi = kr\cos\theta$. Assuming the random distribution of the incident angle which distributes equally likely in all the possible directions, the ensemble averages in the equation become [7][8][9]

$$\mathrm{E}[\cos(kr\cos\theta)] = \frac{\sin kr}{kr} \quad \text{and} \quad \mathrm{E}[\sin(kr\cos\theta)] = 0. \tag{9.26}$$

Consequently,

$$r_s(t,\tau) = \frac{A^2}{2}\rho_3(kr)\,(\cos\omega\tau + \cos(2\omega t-\omega\tau)) \quad \text{and} \quad \rho_3(kr) = \frac{\sin kr}{kr} \tag{9.27}$$

are derived. Taking the time average for a single period,

$$\rho_s(kr,\tau) = \frac{A^2}{2}\rho_3(kr)\cos\omega\tau \tag{9.28}$$

is obtained, where $\rho_3(kr)$ is called the spatial correlation coefficient (or spatial correlation function of r) in a random sound field for the sound pressure responses [7].

The ensemble average is the average over the ensemble that is made by sampling many pairs of the sound pressure records at equal distance by r in the field. The observed phase difference in a pair of samples may be a random variable that follows the phase relation in the random sound field where the angle of the incidence wave is equally likely in the three-dimensional field, even if there is only a sinusoidal source in a room [9].

9.2.2 Two-dimensional random sound field

Sound fields are not necessarily three dimensional. Suppose a random sound field is a two-dimensional space. The spatial correlation function of r can be expressed as [7]

$$\rho_2(kr,\tau)|_{\tau=0} = \frac{2}{\pi}\int_0^{\pi/2}\cos(kr\cos\theta)d\theta = \mathrm{J}_0(kr). \tag{9.29}$$

Eq. (9.29) can be rewritten in terms of the angular spectrum $W(k)$ such that [9][10]

$$\int_0^{k_0} W(k)\cos kr dk = \mathrm{J}_0(k_0 r), \tag{9.30}$$

where

$$W(k) = \frac{2}{\pi}\frac{1}{\sqrt{k_0^2-k^2}} \quad \text{for } 0<k<k_0, \quad \int_0^{k_0} W(k)dk = 1, \tag{9.31}$$

and $\mathrm{J}_0(x)$ is called the 0-th order Bessel function [11].

9.2.3 Cross-correlation function in transient state

The cross-correlation function is defined for the random sound field at the steady state. The correlation function can be extended into the transient state by using the impulse response [12][13]. Sound field responses can be formulated by using the impulse responses from the source to the receiving positions of interest such that

$$p_1(t) = n * h_1(t) \quad \text{and} \quad p_2(t-\tau) = n * h_2(t-\tau), \tag{9.32}$$

where p_1 and p_2 represent the sound field responses at the receiving positions 1 and 2, respectively, similarly $h_1(t)$ and $h_2(t)$ denote the corresponding pair of impulse responses, and $n(t)$ gives the signal of the sound source. Suppose a pair of transient responses to a white noise source from the onset to the steady state. Taking the product and the ensemble average,

$$\rho(t,\tau) = N \int_0^t h_1(t')h_2(t'-\tau)dt' \quad \text{where } \mathrm{E}[n(u)n(v)] = N\delta(u-v). \tag{9.33}$$

The cross-correlation function expressed by Eq. (9.33) is a function of t and τ for representing the transient property including the onset, ongoing, and offset responses, in addition to the steady-state response to the white noise.

The left panel of Fig. 9.5 displays an example of a pair of impulse responses.

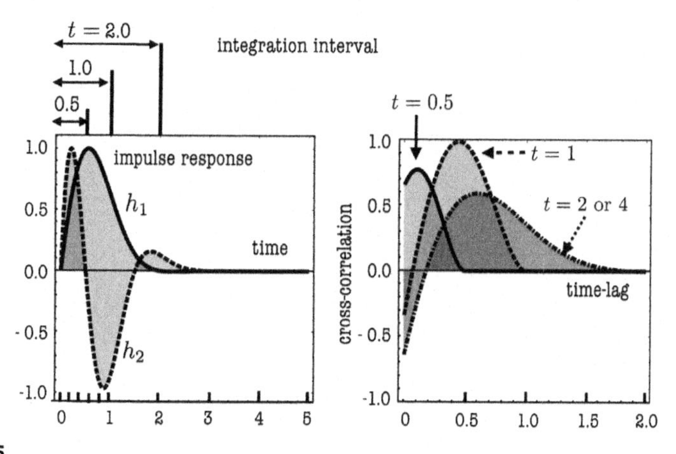

FIGURE 9.5

Sample of impulse response pair (left panel) and examples of transient cross-correlation functions for onset portions (right panel).

The right panel illustrates examples of transient cross-correlation functions for onset portions of the responses to a white noise source according to Eq. (9.33) using the pair of impulse responses, as shown in Fig. 9.5. The transient cross-correlation function converges to that at the steady state, but it may indicate a prominent peak corresponding to the time-lag between the pair of signals at the initial stage (or the duration of integration can be taken at the initial portion of transient signals). The temporal dependence of the correlation function from the initial (onset) to the steady state partly explains why a wideband nonstationary source such as speech material can be localized even in the reverberant space. Onset or transient portions of signals may be helpful in localizing sound sources under reverberant conditions [13].

9.3 Symmetrical sound field in a rectangular room

9.3.1 Spatial correlation for a random sound field in a rectangular room

Suppose that a pair of microphones with spacing of r (m) and a band-limited white noise source to a narrow band (such as 1/3-octave band) are located in a reverberant space. Fig. 9.6 gives an example of the cross-correlation coefficients at the steady state in a rectangular reverberation room [14].

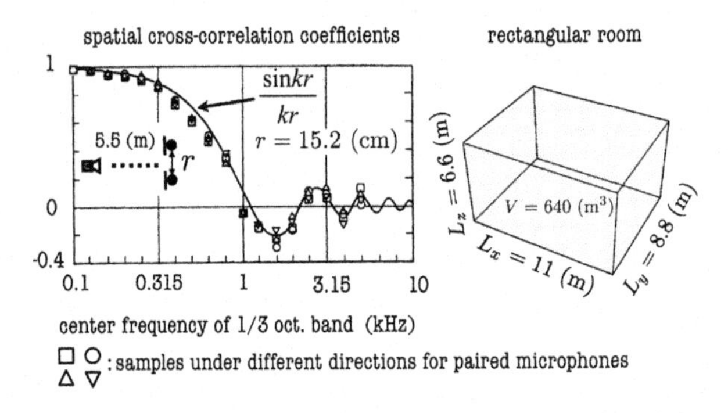

FIGURE 9.6

Sample of spatial cross-correlation coefficients in a rectangular reverberation room where a 1/3-octave band noise source is located from Fig. 1 [14].

The ensemble average can be replaced by taking a long-term time average for a fixed measuring pair in a reverberation field. The cross-correlation coefficients of the sound pressure responses mostly follow the sinc function.

9.3.2 Symmetrical sound field in a rectangular room

A random sound field can be made even in a rectangular room from the perspective of the spatial correlation. On the other hand, the symmetrical nature of a rectangular room can be seen even in the random sound field in a rectangular room. Fig. 9.7 displays examples of measured and calculated spatial correlation coefficients for the symmetrical sound fields in the rectangular reverberant room where a 1/3-octave band noise source is located [15][16]. The sound field is symmetrical with respect to $y = L_y/2$, even in the random sound field excited by a random noise source, as shown in Fig. 9.7. The deviation from the sinc function is due to the symmetry of the random sound field in the rectangular room.

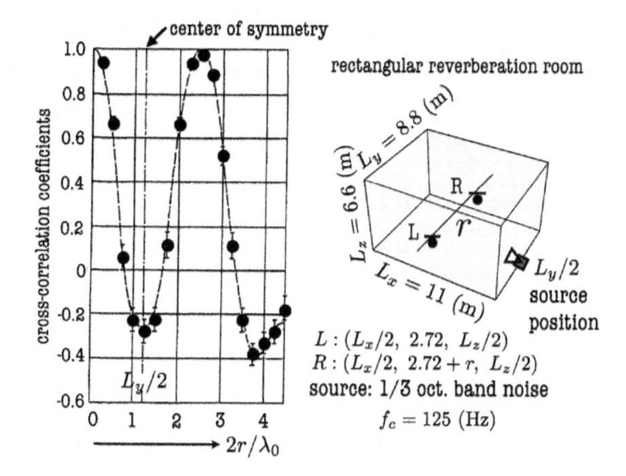

FIGURE 9.7

Spatial correlation coefficients for a random sound field in a rectangular reverberation room by a 1/3–octave band noise source at $(0,\ L_y/2,\ 0)$. Cross-correlation coefficients are taken along the line $(L_x/2,\ 2.72 + r,\ L_z/2)$ with observational error range, where the dashed line shows the numerical calculation from Fig. 6 [16], and λ_0 (m) denotes the wavelength of the center frequency.

9.3.3 Symmetrical binaural sound field in a rectangular room

Listening to a pair of symmetrical signals corresponds to diotic listening because no binaural differences are heard. Recording by a pair of microphones or by a dummy head (KEMAR) [3] in the symmetrical reverberation field mostly produces no binaural differences. Fig. 9.8 displays examples of inter-aural cross-correlation coefficients for symmetrical sound field created in a rectangular reverberation room excited by a 1/3-octave-band noise source [17].

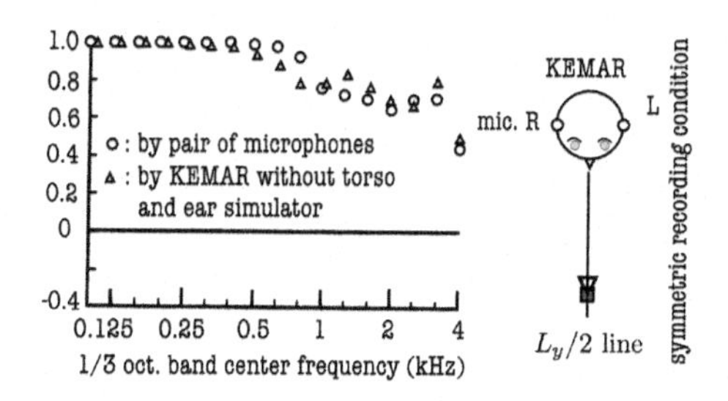

FIGURE 9.8

Inter-aural cross-correlation coefficients for KEMAR in a symmetrical reverberation field.

The sound field is ideally symmetrical, and thus the correlation is expected to be unity. The open circles show the correlation coefficients between the sound pressure responses measured by a pair of microphones whose interval is 15.2 (cm). The symmetrical sound field is mostly created as long as it is seen in the cross-correlation coefficients. The results plotted by the open triangles for KEMAR (excluding the torso and ear simulators) also preserve the symmetry for the most part. The reference [17] mentions that the symmetry might break down around 3 (kHz) when the ear simulators are attached, if there can be differences in the resonance responses between the pair of ear simulators.

9.4 Spatial correlation in random sound fields and subjective diffuseness

9.4.1 Subjective diffuseness of random sound fields

The spatial impression of a sound field is an important attribute of room acoustics from a perceptual perspective [6][18][19][20]. A sound field can be expressed as a superposition of the direct sound and its reverberation from the physical standpoint of room acoustics, as described in Chapters 7 and 8. The direct sound conveys the sound source information, while the reverberation represents the spatial properties of the room in which the sound wave propagates. A representative spatial impression under binaural listening to the reverberation would be subjective diffuseness [21][22][23][24]. The spatial correlation coefficient between an inter-aural pair of signals received in a reverberation field represents the subjective diffuseness in the reverberation field. Subjective diffuseness increases as the correlation coefficients decreases [21]. In particular, the subjective diffuseness is rapidly rendered as the correlation goes down from unity. It recalls that the variance of the correlation coefficients rapidly decreases as the correlation comes close to unity [24]. There is no difference between the inter-aural pair, when the inter-aural correlation is unity, when there is no level difference in the pair. No subjective diffuseness is expected in the symmetrical sound field stated in the previous subsection, while the subjective diffuseness might be quite sensitive to the decrease in the correlation from unity because of some asymmetry.

9.4.2 Direct sound and spatial correlation

Suppose a random sound source in a room. The direct sound makes a spatial correlation function

$$\rho_d = \cos(kr\cos\theta), \tag{9.34}$$

according to Eq. (9.23). The direct sound can be uncorrelated with the delayed sound or reverberation. The spatial correlation function can be defined as a superposition of corresponding correlation functions. The superposition is formulated in terms of the energy ratio of the direct sound and reverberation at a receiving position such that

$$K = \frac{P_0}{4\pi x^2} / \frac{4P_0}{\alpha S} = \frac{\alpha S}{16\pi x^2}, \tag{9.35}$$

following Eq. (7.3), where P_0 (W) denotes the sound power output of the source, x (m) gives the distance from the source, α shows the averaged sound absorption coefficient of the surface of the space, and S (m^2) is the area of the surface. The energy ratio implies that the spatial correlation comes close to the correlation function for the direct sound as the distance from the source decreases, while the correlation function follows the sinc function for the reverberation when the distance increases.

The effect of the direct sound on the spatial correlation function also depends on the directivity of the sound source. In a limit case when the direct sound can be removed, the correlation functions mostly follow the sinc function, almost independently of the distance from the source [8].

9.4.3 Inter-aural correlation in random sound field

The spatial correlation in random sound fields follows a sinc function of kr, where r (m) denotes the distance between the pair of receiving positions and k (rad/m) is the wave number for the central frequency of the narrowband random signal. The inter-aural correlation for a listener may be different from that for a pair of receiving microphones because of the scattering effects by the subject's head and body on the correlation. Actually the minimum audible field could be a little lower than the minimum audible pressure because of the scattering effect [25]. Interestingly, the inter-aural correlation coefficients, mostly follow the sinc function, but $r \cong 32.8$ (cm), as shown in Fig. 9.9 where the cross-correlation coefficients are measured by using the KEMAR dummy head [3] in the rectangular reverberation room [14].

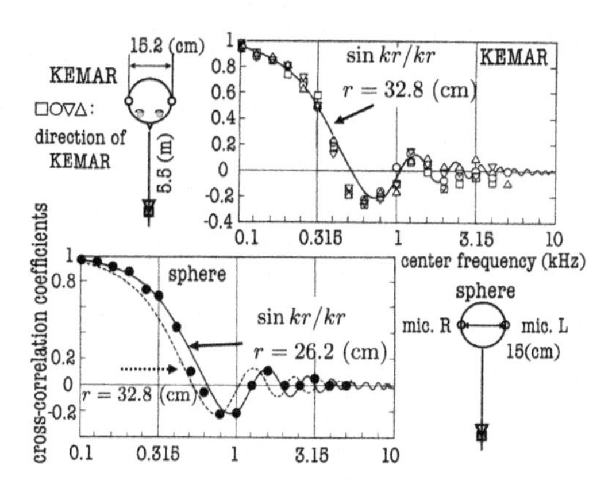

FIGURE 9.9

Inter-aural correlation coefficients in a reverberant sound field measured by using the KEMAR dummy head (upper) and a spherical scatterer (lower) from Fig. 2 [14].

The equivalent inter-aural length in the reverberant sound field is longer than that for the free field [26][27].

The lower panel of Fig. 9.9 displays the spatial correlation by using a sphere instead of KEMAR in the reverberation room. The diameter of the sphere is 15 (cm), which is almost equal to the physical distance between the right and left ears of KEMAR. The spatial correlation coefficients follow the sinc function, but $r \cong 26.2$ (cm), which is understood as the equivalent inter-aural length in the random sound field and is longer than that for the free field [27][28]. The equivalent inter-aural lengths represent the scattering effects of KEMAR and a sphere on the inter-aural correlations.

9.4.4 Inter-aural correlation in a mixed sound field

A mixed field of direct sound and reverberation is rendered in a small room where the reverberation time is not too long [29]. Fig. 9.10 displays an example of spatial correlation coefficients in a mixed sound field rendered in a small office by a single loudspeaker, with volume 55.8 (m^3) and reverberation time around 0.21 (s).

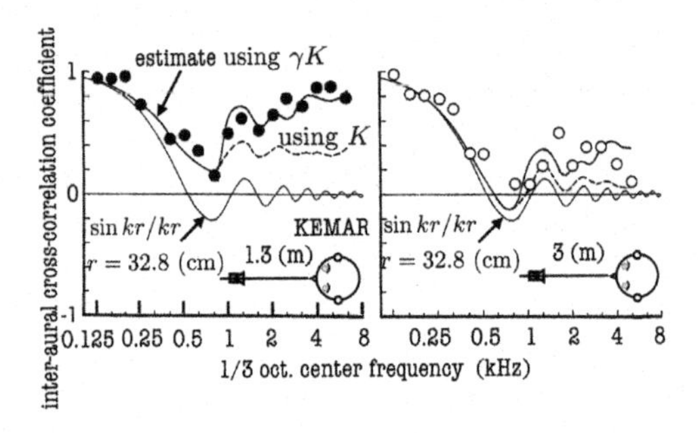

FIGURE 9.10

Samples of spatial correlation coefficients in a small room. Sound source distance is 1.3 (m) (left panel) and 3 (m) (right panel).

The reverberation time and the direct to reverberant energy ratio estimated by Eq. (9.35) are shown in Fig. 9.11.

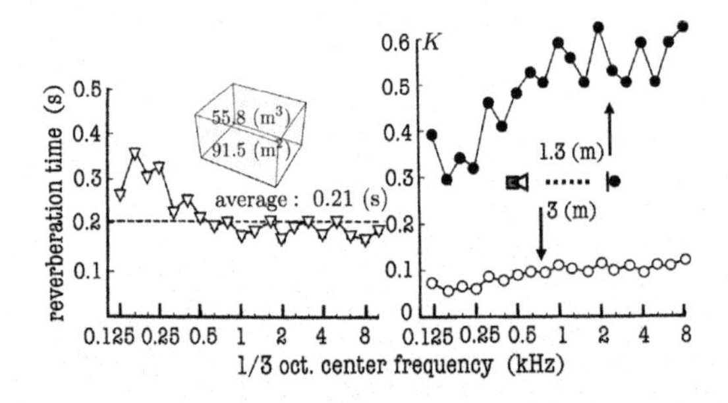

FIGURE 9.11

Left panel: Reverberation time of the small room used for Fig. 9.10. Right panel: Energy ratio of direct sound to reverberation at the receiving positions.

Inter-aural correlation coefficients can be expressed as a superposition of those for the direct sound and reverberation in a mixed field such that

$$\rho = \frac{1}{K+1}(K\rho_d + \rho_3) = \frac{1}{1+1/K}(\rho_d + \rho_3/K), \tag{9.36}$$

where $\rho_d = 1$ for the direct sound under the condition that the listener (or KEMAR) faces the loudspeaker, and ρ_3 means the sinc function for the reverberation. Taking the directivity of the sound source into account, the energy ratio becomes

$$\tilde{K} = \gamma K, \tag{9.37}$$

where γ denotes the directivity of the source. Introducing the directivity of a source into the energy ratio, then the inter-aural correlation coefficients can be rewritten as [30]

$$\rho = \frac{1}{1+1/\tilde{K}}(\rho_d + \rho_3/\tilde{K}). \tag{9.38}$$

Fig. 9.10 confirms that inter-aural correlation coefficients mostly follow a superposition of the correlations corresponding to those of the direct sound and reverberation by taking into account the directivity of the loudspeaker system used in the experiments [29]. The results imply that spatial impression, for example the subjective diffuseness, might not be rendered in a small room by a single channel of sound source mostly because of the directivity of the source. However, the question arises of why the subjective diffuseness is not clearly sensed in small rooms, even if the correlation function is close to the sinc function in the sound field as the distance from the loudspeaker increases [31].

The inter-aural correlations are expressed by a pair of impulse responses when a sound source radiates white (or band-limited) noise. The subjective diffuseness of sound fields in terms of the inter-aural correlations tacitly assumes the sound source signal is a random noise. This assumption is required in the formulation of the correlation functions in the mixed sound field composed of the direct sound and reverberation. Speech or musical sound, in an everyday situation of listening to sound in rooms, is transient in the time domain and is different from a wideband random noise. The transient correlation quite likely takes high peaks as implied in Section 9.2.3. Peaks that occur very frequently in the transient correlation functions may spatially localize a sound source even under highly reverberant conditions. It would be intuitively understood by recalling that a source position of speech signals can be easily detected in a reverberation room, while it may be quite unlikely for a sinusoidal or random-noise source.

Interestingly, D_{30} might be a candidate of estimating a portion of the direct sound for inter-aural correlations from a perceptual standpoint, although D_{30} is originally defined as representing the monophonic information of sound field, as described in Section 7.1.2 [31][32][33][34]. Early reflections within 30 (ms) may be fused into the direct sound rather than the distinct echoes or reverberation of the direct sound. Substituting $D_{30} = K/(1+K)$ for Eq. (9.36), the inter-aural correlations of mixed fields can be formulated as [31]

$$\rho_{mix} = D_{30}\rho_d + (1 - D_{30})\rho_3, \tag{9.39}$$

where the subjective diffuseness may be sufficiently lower, as if it might represent the correlations in an anechoic room. Low subjective diffuseness seems to be consistent with our daily experiences in small rooms.

9.4.5 Inter-aural correlation of a two-channel reproduced sound field

Reproducing the inter-aural correlations of a reverberation field seems difficult by a single loudspeaker in a room because of the dominance of the direct sound including early reflections in the room. An anechoic room is a limit case where there is only the direct sound from a single loudspeaker without reflected sound. The inter-aural correlation is expressed as Eq. (9.34) in an anechoic room, which localizes an image source without the subjective diffuseness of sound fields.

Suppose a pair of loudspeakers instead of a single one that radiates uncorrelated reverberation. The uncorrelated pair of reverberations can be recorded by a clearly separated pair of microphones far from the sound source in the original field. Examples of the reproduced inter-aural correlation coefficients are shown in the left panel of Fig. 9.12 [14] under two-channel reproduction conditions in an anechoic room [35].

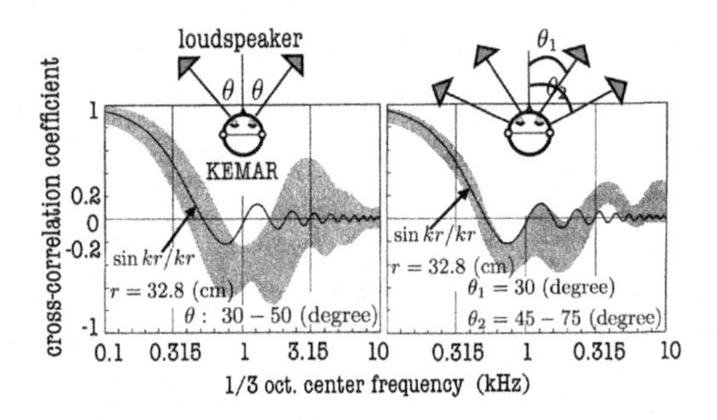

FIGURE 9.12

Sample of inter-aural cross-correlation coefficients reproduced in an anechoic room by two-channel (left) and four-channel (right) reproduction from Figs. 4 and 6 [14], where the KEMAR dummy head is used and sound fields are rendered by 1/3-octave band noise, and the range of data are shown here.

Even in the free field, the reproduced inter-aural correlation coefficients by two-channel stereophony more or less follow the sinc function of kr ($r = 32.8$ (cm)) that is the desired inter-aural correlation coefficients for reproduction of the original reverberation. Reproduced inter-aural correlation coefficients would partly explain why the conventional two-channel-stereophonic recording is so widely accepted by the consumers.

Looking at the details by comparison with the results of four-channel stereophony, however, negatively or positively deviated correlations from the sinc function can be seen in the middle- or high-frequency range for two-channel reproduction. These troughs or peaks in the correlation imply an unnatural separation of sound images into the left or right loudspeakers. The correlations rendered by four-channel reproduction (right panel) follow more closely the sinc function than those by the two-channel reproduction.

9.5 Reproduction of inter-aural correlations for mixed sound fields

9.5.1 Mixing of direct sound and reverberation

The sound field can be expressed as a superposition of the direct sound and reverberation. Recalling the definition of D_{30}, the direct sound can be extended to include the early reflections within 30 (ms) from the subjective and signal theory perspectives. The spatial correlation coefficients in the original reverberation field, such as a concert hall, can be formulated as

$$R_0 = \frac{K + \rho_3}{1 + K}, \tag{9.40}$$

according to Eq. (9.36). Reproducing the cross-correlation coefficients of the original field in other listening spaces would be of basic interest for sound recording and reproduction from a subjective standpoint. Suppose that $\rho_3 \cong 0$ at the pair of microphones far from the sound source under the recording condition so that mostly the reverberant sound $y(t)$ might be recorded. Take another microphone at a position very close to the sound source for recording the direct sound $x(t)$. The total sound to be recorded is a superposition of the direct sound and reverberation with the mixing ratio $\beta = \overline{x^2}/\overline{y^2}$ as a parameter where $\overline{*}$ denotes the long-term average.

Suppose a pair of recording sounds

$$s_L(t) = x(t) + y_L(t) \text{ and } s_R(t) = x(t) + y_R(t), \tag{9.41}$$

where the pair of y_L and y_R represents the reverberation, and $x(t)$ denotes the direct sound. Assume that the listening space is a free field. Simplifying the reproduced sounds as

$$\begin{aligned} u_L(t) &= x(t) + x(t-\tau) + y_L(t) + y_R(t-\tau) \\ u_R(t) &= x(t) + x(t-\tau) + y_R(t) + y_L(t-\tau), \end{aligned} \tag{9.42}$$

where τ (s) denotes the time delay between the pair of receiving points and the magnitude difference between the listening pair is neglected. The cross-correlation coefficient between the pair of listening (or reproduced) sounds can be written as

$$R_s = \frac{\overline{u_L(t) \cdot u_R(t)}}{\sqrt{\overline{u_L^2(t)}} \cdot \sqrt{\overline{u_R^2(t)}}} \cong \frac{\beta(1+\rho) + \rho}{\beta(1+\rho) + 1}, \tag{9.43}$$

assuming that

$$\rho = \rho_x = \frac{\overline{x(t) \cdot x(t-\tau)}}{\sqrt{\overline{x^2(t)}} \cdot \sqrt{\overline{x^2(t-\tau)}}} \cong \rho_y, \tag{9.44}$$

where

$$\overline{y_L^2(t)} = \overline{y_R^2(t)} = \overline{y^2(t)} \text{ and } \overline{x^2(t)} = \overline{x^2(t-\tau)} \tag{9.45}$$

and

$$\overline{x(t) \cdot y_L(t)} = \overline{x(t) \cdot y_R(t)} = \overline{y_L(t) \cdot y_R(t)} = 0. \tag{9.46}$$

The mixing parameter β for recording sound is desired so that R_0 might be equal to R_s.

Fig. 9.12 shows the spatial correlations in a two-channel reproduced field by using a pair of 1/3–octave random noise sources that correspond to the mixing ratio $\beta = 0$. Following Eq. (9.43), the correlation coefficients R_s can be written as

$$R_s \cong \rho \tag{9.47}$$

for $\beta = 0$, where ρ is given by Eq. (9.44) and corresponds to the auto-correlation function of the recorded signal $x(t)$, where the time lag is given by the spacing between the two receiving positions. The deviations from the sinc function in Fig. 9.12 may reflect the auto-correlation functions of the random noise signals at the receiving positions.

9.5.2 Reproduction of spatial correlation

Mixing control is an attractive factor in sound recording and reproduction, such that the spatial correlations under the recording condition might be preserved even in a listening room. Setting

$$R_0 = \frac{K + \rho_3}{1 + K} = R_s = \frac{\beta(1+\rho) + \rho}{\beta(1+\rho) + 1}, \tag{9.48}$$

then

$$\beta = \frac{R_0 - \rho}{(1+\rho)(1-R_0)} \tag{9.49}$$

is obtained as a function of ρ. The mixing ratio β depends on the auto-correlation function of the recorded signal $x(t)$ given by Eq. (9.44). The ratio can be well defined so that $R_0 \cong R_s$, assuming that $R_0 > \rho$ and $\rho \neq -1$.

The correlation in the recording field, R_0, however, could be close to 0 following the sinc function. On the other hand ρ might not always be around 0 in every 1/3–octave band following the auto-correlation function of the random noise. Consequently, the mixing ratio is negative for $R_0 < \rho$ when the ratio is not well defined. In particular, $\beta = -\rho/(1+\rho)$ when $R_0 = 0$. Assuming ρ may represent the correlation coefficient in the two-channel reproduced field, the negative deviation ($\rho < 0$) from the sinc function might be compensated by the mixing ratio, while the positive deviation ($\rho > 0$) may not be the case. The control of inter-aural correlations in the reproduced field may be extended from the listening points into the area by multichannel reproduction so that the wave front of the sound might be resynthesized, as reported in the references [36][37][38].

9.5.3 Decreasing positive correlation of reproduced sound by anti-phase mixing

Anti-phase correlation components are necessary to decrease the correlation [39]. Suppose a pair of recording sounds

$$\begin{aligned} u_L(t) &= x(t) + x(t-\tau) + y_L(t) + y_R(t-\tau) + y_c(t) - y_c(t-\tau) \\ u_R(t) &= x(t) + x(t-\tau) + y_R(t) + y_L(t-\tau) - y_c(t) + y_c(t-\tau), \end{aligned} \tag{9.50}$$

where $y_c(t)$ is recorded at another position in between the pair of microphones for the reverberation, assuming

$$\overline{y_L(t) \cdot y_c(t)} = \overline{y_R(t) \cdot y_c(t)} = 0. \tag{9.51}$$

Following a similar procedure to Eq. (9.43), the correlation coefficients in the reproduced sounds can be formulated such that

$$R_{ss} \cong \frac{\beta(1+\rho) + \rho - \alpha(1-\rho)}{\beta(1+\rho) + 1 + \alpha(1-\rho)} \tag{9.52}$$

as a function of ρ, where

$$\alpha = \overline{y_c^2(t)}/\overline{y^2(t)}. \tag{9.53}$$

Setting

$$R_{ss} = R_0 \tag{9.54}$$

so that the reproduced correlation coefficient might become R_0 by controlling α, as defined in Eq. (9.53). The energy ratio α for the anti-phase reverberation component can be derived as

$$\alpha = \frac{\beta(1+\rho)(1-R_0) + \rho - R_0}{(1-\rho)(1+R_0)} \tag{9.55}$$

when

$$\beta(1+\rho)+1+\alpha(1-\rho)\neq 0. \tag{9.56}$$

Assuming that $R_0 = 0$ and $\rho > 0$ represent the positive deviation in the reproduced correlation coefficient, the anti-phase mixing ratio α can be rewritten as

$$\alpha = \frac{\beta(1+\rho)+\rho}{1-\rho}, \tag{9.57}$$

which may be realizable for $\rho > 0$.

The inter-aural correlation coefficients in a two-channel reproduced field can mostly follow the trend of the sinc function, which the correlations follow in the reverberation field except prominent positive or negative deviations, as shown in Fig. 9.12. The positive or negative deflections might be partly equalized by mixing the direct sound and reverberation according to α or β [39].

An anechoic room and stationary random noise sources are used to investigate inter-aural correlation coefficients in reproduced fields. A difference in the nonstationary nature in the time domain between random noise and speech (or musical sounds) can be taken into account by introducing D_{30} into the energy ratio of direct sound and reverberation, as described in Eq. (9.39) [31]. Under everyday listening conditions, most reverberation in the listening room can be within around $30-50$ (ms) after the direct sound. Such reverberation can be fused into the direct sound if the direct sound can be reinforced, and thus the energy ratio increases. The reproduced inter-aural correlations in an anechoic room may be representative for the reproduced sound in listening rooms, assuming that the subjective energy ratio for D_{30} holds under everyday listening conditions.

9.6 Binaural detection for frequency dependence of inter-aural correlation

The large positive and negative deviations from the sinc function can be found in the reproduced inter-aural correlation coefficients by two-channel loudspeakers. This section develops possibilities in the discrimination of the inter-aural correlations by binaural listening using wideband or narrowband random noises [14][40].

9.6.1 Binaural sensing of differences in frequency characteristics of inter-aural correlation

The frequency characteristics of reproduced inter-aural correlation coefficients convey the spatial impression of the original field. An intriguing question would be whether or not the differences can be sensed in the frequency characteristics of the cross-correlation coefficients [21]. Speech intelligibility or pitch sensation is understood according to sub-band filtering models [41][42]. If the model works for the perception of inter-aural correlations, sensation of the difference in the frequency dependence of the inter-aural correlations is possible.

Fig. 9.13 shows examples of the frequency dependence of the inter-aural correlation coefficients [14].

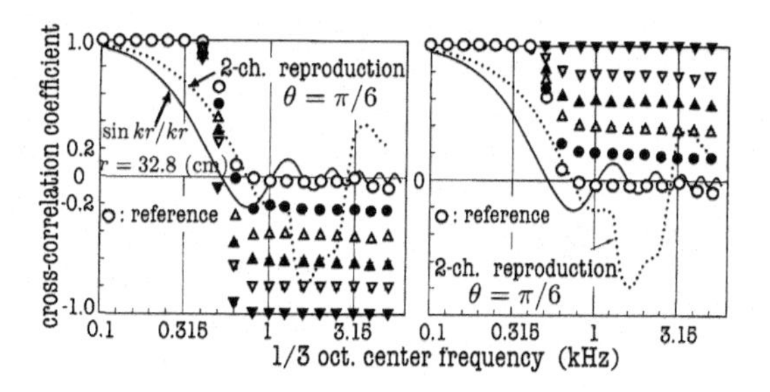

FIGURE 9.13

Examples of frequency dependent cross-correlation coefficients from Fig. 9 [14] where the deflection for inter-aural correlation coefficients under two-channel reproduction is simply reflected.

Broadband pink noise (0.1 – 5 (kHz)) pairs with the cross-correlation coefficients illustrated in Fig. 9.13 are made. The noise pairs have high correlations below 500 (Hz), and either negative (left panel) or positive (right panel) high correlation above 500 (Hz). The characteristics represent, more or less, the typical patterns for the two-channel stereo reproduction, where the sinc function is shown by the solid line for the reference.

Binaural listening experiments are carried out on sensing the differences in spatial impression such as the subjective diffuseness for noise samples with respect to cross-correlation coefficients, as shown in Fig. 9.13. Pairs of noise samples are presented to listeners through headphones under dichotic listening conditions. The experiments were performed with 11 subjects who were not specially trained for binaural listening tests. In each trial a pair of noise bursts were emitted, each with an interval of 2 (s) and a listening level of about 70 (dB). The subjects were asked to answer "yes" or "no" when they did or did not experience a different spatial impression between a pair of noise bursts. The percentage of subjects who answered "yes" during each trial was tabulated.

Fig. 9.14 summarizes the results of sensing the differences from the reference in the cross-correlations under the dichotic listening condition [14]. The reference pair of signals whose cross-correlation coefficients are marked by the open circles roughly represents the trend of the frequency dependence of the sinc function, while the solid triangles ($|\rho| = 0.6$ for higher than 500(Hz)) partly represent the deviation by the two-channel reproduction from the sinc function. The listening scores in Fig. 9.14 imply that the deviations might be noticeable. An unnatural spatial impression might be rendered by conventional two-channel stereophony.

Another example of pairs of wideband random noise signals are studied in reference [40], where cross-correlation coefficients of noise pairs are frequency dependent. The frequency dependence is displayed in Fig. 9.15.

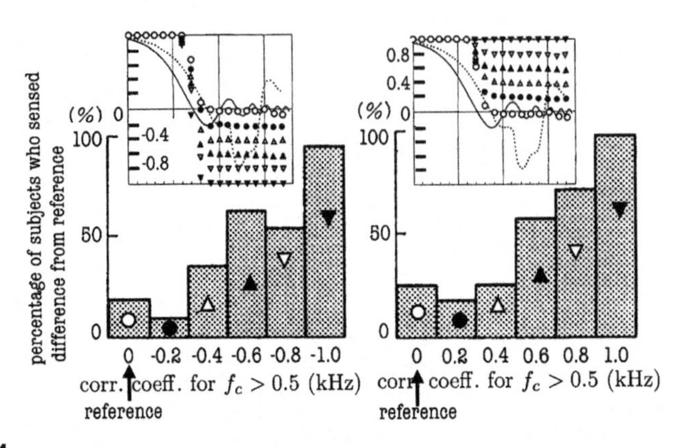

FIGURE 9.14

Percentage of subjects who sensed the difference from the reference in frequency dependence of inter-aural correlation coefficients where the reference is the noise sample shown by open circles: left (right) for negative (positive) correlations for $f_c > 0.5$ (kHz) from Fig. 10 [14].

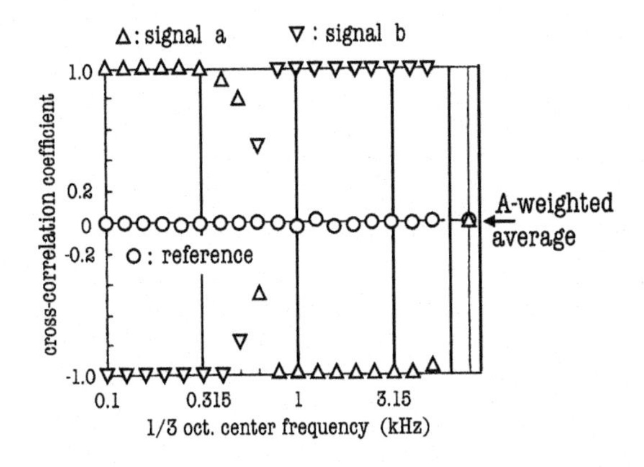

FIGURE 9.15

Samples of frequency characteristics of cross-correlation coefficients of pairs of wideband random noise.

Three types of frequency characteristics are made for pink noise pairs, where A-weighted averages of the three-types are almost equal to 0. The same types of listening tests for Fig. 9.14 were performed. Interestingly, the subjective diffuseness can be rendered only for the reference, while the spatial impression is quite unlikely for other noise pairs. The detection of the difference in the spatial impression is quite easy in the tests. Sub-band filtering is also a fundamental basis for spatial impressions such as the subjective diffuseness, although the effects of the headphones on the perception of spatial impression might not be excluded in the experiments.

9.6.2 Binaural sensing of inter-aural correlation for narrowband noise pairs

The perception of spatial impressions might be on a sub-band basis [21]. An interesting question for two-channel reproduction, would be whether the deviations of reproduced inter-aural correlation from the sinc function can be subjectively perceived for sub-band noise signals. Fig. 9.12 illustrates the "envelope" of the reproduced inter-aural correlation coefficients reproduced by two-channel reproduction, where $30° \leq \theta \leq 50°$. Take the upper and lower limits in the envelope. The same types of listening tests as shown in Fig. 9.13 were performed, but using 1/1−octave-band noise pairs rather than wideband noise pairs. The reference is the sinc function at every frequency band [43].

Fig. 9.16 gives the number of subjects who noticed differences in the cross-correlation coefficients from the sinc function at the lower or higher limits of the envelope.

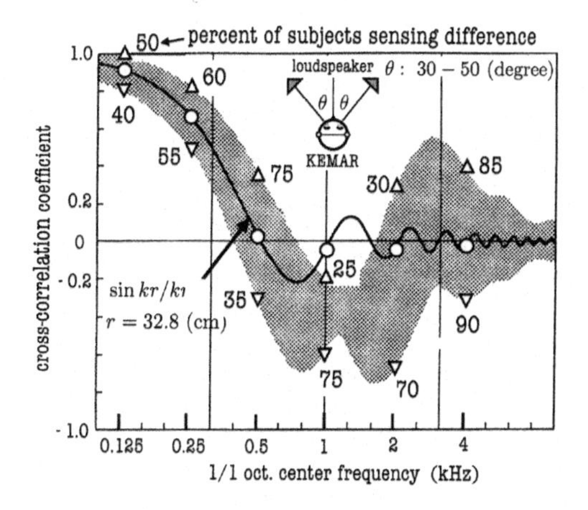

FIGURE 9.16

Envelope of correlation coefficients with percentage scores of number of subjects who noticed differences from the sinc function at every 1/1−octave band where the lower or higher limit of reproduced correlations are taken.

The positive deviation at around 500 (Hz) is possibly perceived, and the negative deviations at frequency higher bands than 500 (Hz) seem quite likely to be noticed. In addition, the positive deviation around 4 (kHz) can likely be sensed.

The results in Fig. 9.16 might explain the unnatural separation (or overseparation) between two-channel loudspeakers in the conventional stereo reproduction. The positive deviation around 500 (Hz) indicates the lack of subjective diffuseness in the two-channel reproduction [31][39]. Signals can be recorded by mixing the direct sound and reverberation. Mixing by three microphones described in the previous section would be a possible way to equalize the unnatural spatial impression in two-channel stereophonic reproduction.

9.7 Precedence effect on sound localization and envelope of signals

9.7.1 Precedence effect

A sound field in a room is composed of direct sound and reverberation. The inter-aural correlation follows the sinc function corresponding to the subjective diffuseness in a room, as the distance increases from the source, when the sound source radiates random noise. However, for speech signals a listener is able to localize the sound source even in the reverberant space. Early reflections within around 30 (ms) of the direct sound enhance the energy of the direct sound. This positive enhancement due to early reflections can be explained by the Haas effect or the precedence effect [18][44][45][46] in terms of the binaural sound localization.

Fig. 9.17 illustrates a schematic of the superposition of the direct sound and a single reflection of the direct sound (representing the early reflections by a single one), where the delay time of the single reflection is indicated by τ (s).

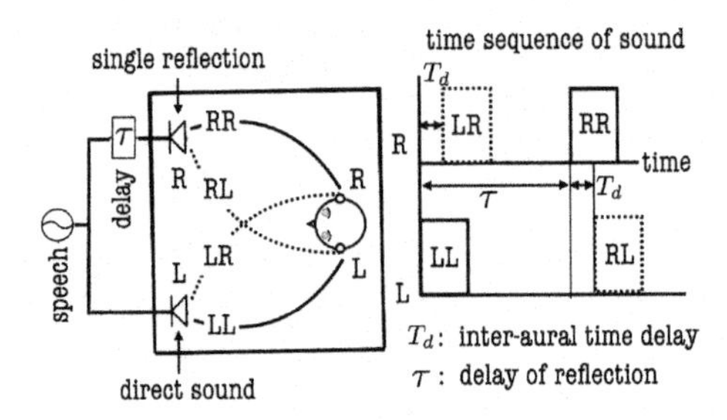

FIGURE 9.17

Schematic of superposition of direct sound and a single echo at binaural listening positions in an anechoic room.

The right panel displays a time sequence of the direct sound followed by a single reflection at both of the ears of a listener. Two types of time delays can be seen: the inter-aural time delay T_d within around 1 (ms) and the delay τ (ms) of the reflection (s). A fused sound image can be made between the two loudspeakers, when the signal delay τ is shorter than around 1(ms) or the maximum of the inter-aural time difference T_d. Interesting is the case when the signal delay is longer than the maximum inter-aural time delay, such as $\tau > 1$ (ms). The sound is localized to the "real sound source" (not an image source) radiating the direct sound corresponding to the left loudspeaker in the figure, as if no reflection (or delayed sound) comes from the right loudspeaker in the left panel. This is called the precedence effect or Haas effect, which masks the reflection with regard to the binaural sound localization, although the timber of the superposed sound may change from the original value without the reflection.

When the signal delay τ becomes longer than around 30 (ms), the echo is split and can be clearly perceived by the listener. The perception of the split echo is noticed by the listener as a reflection coming from the right loudspeaker. Consequently, two sound sources (left and right loudspeakers) are separately localized corresponding to the direct and delayed sound, respectively, without fusion into the direct one.

Fig. 9.18 is an example showing the precedence effect on speech samples that are composed of the direct sound followed by a single reflection [47].

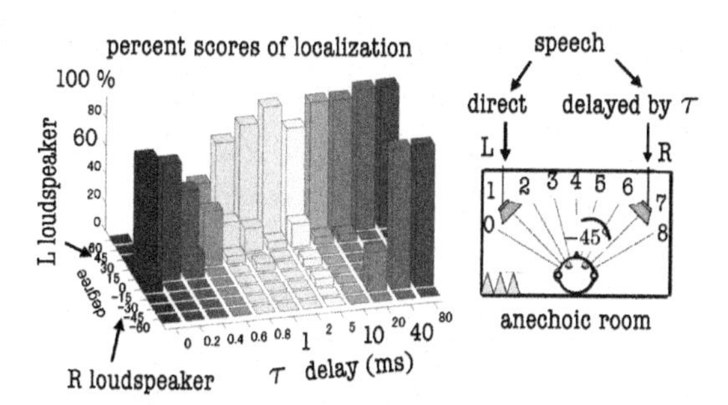

FIGURE 9.18

Example of precedence effects on speech materials composed of direct sound followed a single reflection.

The illustration explains the precedence effect on the direct sound and a single reflection with the delay time τ for a speech material. When the delay time τ is longer than around 30 (ms), the delayed sound is separately perceived as an echo of the direct sound.

9.7.2 Broadband noise and the precedence effect

The precedence effect on the direct sound and a single reflection with a delay time τ for a speech sample was introduced in the previous subsection. When the delay time τ is longer than around 30 (ms), the delayed speech is distinctly perceived as an echo of the direct speech. A remarkable difference can be seen for the effects on broadband noise. A direct and delayed random noise pair renders the subjective diffuseness instead of a separated single echo [48][49].

Fig. 9.19 gives the differences in the effects of a single echo with a time delay on the sound localization for wideband noise (left) or speech material [48][49]. Three subjects were asked to identify the directions of a single number (1 – 11) or multiple numbers in order to display the range of the localization area when a broad sound image is made. The left panel shows the results for wideband noise (< 6 (kHz)). The sound image broadens as the delay time exceeds around 0.8 (ms), and spreads over the inside of the two loudspeakers when the delay time reaches around 20 (ms). The subjective diffuseness is rendered when the delay is longer than 20 (ms), and no sound image splits into direct sound and reflection.

In contrast, the right panel displays the case for the speech material. The sound image moves to the left side as the delay time increases, and the left-side loudspeaker is clearly identified as the delay time is even longer than about 0.6 (ms). The localization holds as long as the delay time does not exceed 4 (ms). The sound image

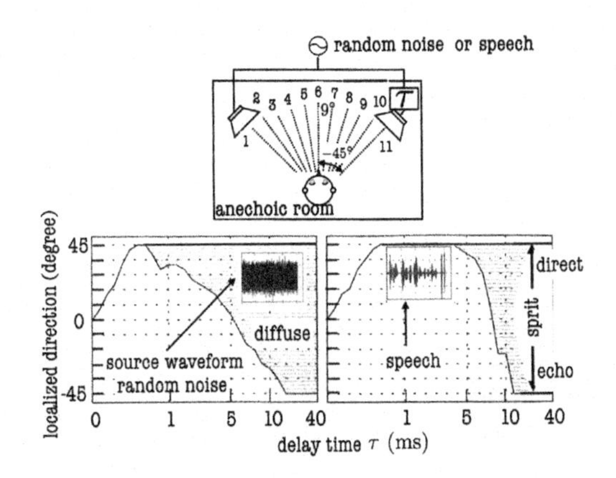

FIGURE 9.19

Differences in precedence effects on sound localization for wideband random noise (left) and a speech sample (right).

broadens at the delay time of about 4 (ms), and the image splits into the direct and delayed sound at around 10 (ms) of the delay time with neither fusion nor subjective diffuseness.

The difference between the wideband noise and speech material is clear. The precedence effect is remarkable for the speech material, while the random noise is not when $\tau > 1$ (ms). The echo due to the delayed sound is clearly heard for the speech after the precedence effect, while no split echo is perceived for the random noise.

Looking at the samples of waveforms as shown in Fig. 9.19, the envelopes of the waveforms might explain the differences in the precedence effect [47][48][49]. The signal or nonstationary dynamics in the time domain may be important in the precedence effect or in the detection of a split echo. The difference in the precedence effects of speech and random noise materials motivated us to introduce Eq. (9.39) into the subjective diffuseness for musical signals composed of many transient portions [31]. In the next subsection a modulated random noise is taken as a model of speech waveforms with respect to the precedence effect including detection of the echo.

9.7.3 Modulation of broadband random noise and the precedence effect

One way to approach a difference in the signal dynamics of a speech material from wideband random noise is to modulate a wideband random noise by envelope of the speech waveform. The modulation of a random noise can be formulated as [49]

$$y(t) = ((1-m) + m \cdot s_e(t)) \cdot x(t), \tag{9.58}$$

where $x(t)$ denotes the wideband random noise, $s_e(t)$ is the Hilbert envelope defined as the magnitude of the analytic signal representation of the speech signal $s(t)$, and m indicates the modulation index.

The same type of experiments on the precedence effect were carried out for modulated noise as those for random noise or the speech sample. The left panel of Fig. 9.20 is an example of the precedence effect on the modulated wideband noise.

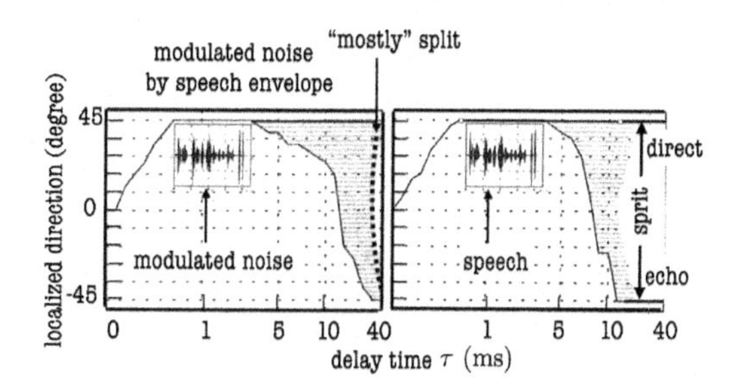

FIGURE 9.20

Precedence effect of modulated random noise. Left: Modulated ($m = 1$) wideband noise by envelope of the entire speech waveform. Right: Speech for reference.

The envelope of the modulated noise is the same as that for the speech waveform, although the modulated noise is almost completely unintelligible, independent of the delay time [41]. The localization of the leading sound (modulated noise) clearly holds and is almost the same as that for the speech; the sound is localized even when the delay time reaches 4 (ms). Sound source localization under the precedence effect might be mainly due to the envelope or the signal dynamics rather than the fine structure of the waveform [47].

In addition to the localization, the split image into the leading (direct) and delayed (reflection) sound can be rendered even for the modulated noise; however, the time delay at which the splitting starts is different between the modulated noise and the speech sample. The splitting of the echo might be mainly due to the effect of the envelope [47][48][50]. The references [48][50] suggest that the delay time of a detectable splitted echo increases as the modulation index m of a speech sample decreases.

The delay time at which echo splitting starts indicates the break down of sound image broadening or the subjective diffuseness. No sound image broadening reaching the subjective diffuseness can be expected for speech samples. As the sound image becomes broad by increasing the delay time, the image splitting starts simultaneously. In contrast, the subjective diffuseness might be possible for the modulated noise, although the spatial impression breaks in a different way from wideband noise without envelopes, as the delay time increases. The subjective diffuseness can be due to the randomness in the fine structure of the waveform, as long as the signal dynamics represented by the envelope does not yield the splitting echo. According to the representation of the modulated noise given by Eq. (9.58), the modulated noise can be rewritten as

$$y(t) = s_e(t) \cdot x(t) = y_e(t) \cdot x_c(t) \text{ and } y_e(t) = s_e(t) \cdot x_e(t), \tag{9.59}$$

where $y_e(t) \neq s_e(t)$. The expression in Eq. (9.59) partly explains the difference in the echo detection or subjective diffuseness between the modulated noise and speech. The Hilbert envelope of the modulated noise y_e is different from the speech envelope s_e, and the envelope of the random noise x_e may again be random noise.

9.7.4 Envelope correlation and precedence effect

The precedence effect for speech samples or modulated random noise suggests the relationship between the precedence effect and envelope auto-correlation function [48]. Suppose the auto-correlation function for squared modulation envelopes, such as

$$r_e(\tau) = (1 - m) + m \cdot \cos \omega_e(\tau), \tag{9.60}$$

where m denotes the modulation index. When m approaches 0, the envelope correlation comes close to being flat without the fluctuation in the time domain. In contrast, as m approaches unity, the envelope correlation function does not have its "direct current" components. The envelope correlation function of the squared waveform of the wideband random noise can be assumed by $m \cong 0$, such that $r_{e_{\text{noise}}}(\tau) \cong 1$. A flat correlation of the squared envelope with a random carrier means that the subjective diffuseness is quite likely, in contrast to the precedence effect and splitting echoes which are quite unlikely.

The auto-correlation of the envelope of a squared speech waveform is composed of lower frequency components than those of the carrier. Suppose the frequency components are spread over $\pm\Delta\omega$. The frequency averaged auto-correlation function can be written as

$$\overline{r}_{e_{\text{speech}}}(\tau) = (1 - \overline{m}) + \overline{m} \cdot \frac{1}{2\Delta\omega} \int_{-\Delta\omega}^{\Delta\omega} \cos \omega_e \tau d\omega_e \cong \frac{\sin \Delta\omega t}{\Delta\omega t}, \tag{9.61}$$

assuming that the average of the modulation index $\overline{m}$ might be close to unity. The formulation of Eq. (9.61) gives an estimation of the time delay when the echo is perceived. Take the time τ_{Min} that sets the auto-correlation of the envelope to zero, and assuming a possible frequency range of the squared envelope to be about 20 (Hz), then $\tau_{Min} \cong 25$ (ms) is obtained similarly to Eq. (7.27). This estimation gives a possible interpretation for the detection of the splitting echo, and implies a schematic of the precedence effect.

Suppose a time windowing function to be the auto-correlation function for the squared envelope. Following the assumed windowing function, the wideband noise with a random carrier of $m \cong 0$ would have an excessive auto-correlation time for the squared envelope to discriminate the echo after a short interval from the direct sound. The fused image without splitting the sound renders the subjective diffuseness due to the pair of independent noise sounds, as described in Section 9.4.1

In contrast, the auto-correlation function of a squared speech envelope would indicate an appropriate windowing function that makes the splitting echo occur within $20 - 30$ (ms) after the direct sound. The echoes coming within a shorter time interval than the window length may fuse into the direct sound. The fusion produces a single image without splitting due to the correlated envelopes of the direct and delayed waveforms. The precedence effects imply that the auto-correlation function of the squared modulation envelope of sound might give a possible candidate of time-window range of the binaural listening that renders sound localization, subjective diffuseness, and the detection of the echoes.

9.8 Exercises

1. Confirm the pair of Eqs. (9.9) and (9.10).

2. Obtain Eqs. (9.14) and (9.15) [5].

3. Derive Eq. (9.28).

4. Confirm Eq. (9.33) [12].

5. Reconfirm Eq. (9.35).

6. Express the random variable Z

$$Z = \frac{(X - Y)^2}{(X + Y)^2}, \quad \text{where } \mathrm{E}[X^2] = \mathrm{E}[Y^2] = 1, \tag{9.62}$$

by using the cross-correlation coefficients between X and Y. The relationship was used to measure the cross-correlation coefficients in the "analog" days. The cross-correlation would be intuitively understood by the sum and difference for the corresponding pair of random variables.

7. Derive Eq. (9.43), where ρ denotes the auto-correlation of the source signal.

8. Show the auto-correlation function of band-limited white noise whose bandwidth is $\Delta\omega$. Compare of the auto-correlation function of the time lag for the random noise and the spatial auto-correlation function ρ_3 of r [7][9][15][16]. In particular, if some readers are interested in the theoretical background of the statistical acoustics for random sound fields, reference [9] would be a very nice and high level textbook.

9. Consider the reproduced inter-aural correlation coefficients by symmetrical front-back four-channel reproduction [14]. Enhancement of subjective diffuseness would not be feasible even via symmetrical four-channel reproduction because of the symmetry.

10. Explain the assumption for Eq. (9.39) according to the precedence effect on speech-like signals [31].

11. Explain the terminologies listed below:
(1) binaural magnitude ratio (2) binaural phase difference (3) head-related transfer function
(4) binaural filters (5) cross-talk cancelation (6) cross-correlation function
(7) wave number (8) speed of sound (9) spatial correlation coefficient
(10) spatial correlation function (11) Bessel function (12) two-dimensional reverberation field
(13) angular spectrum (14) correlation function at transient state (15) temporal change of correlation
(16) spatial correlation of random sound field (17) cross-correlation coefficient for symmetrical sound field (18) diotic listening
(19) subjective diffuseness (20) energy ratio of direct sound to reverberation
(21) equivalent inter-aural length
(22) inter-aural correlation coefficient in mixed field (23) directivity of sound source (24) precedence effect
(25) Hilbert envelope (26) modulation index (27) auto-correlation function of squared envelope

References

[1] E.A.G. Shaw, Transformation of sound pressure level from the free field to the ear drum in the horizontal plane, J. Acoust. Soc. Am. 56 (6) (1974) 1848–1861.
[2] P. Damaske, Head-related two-channel stereophony with loudspeaker reproduction, J. Acoust. Soc. Am. 50 (4) (1971) 1109–1115.
[3] B.D. Burkhard, R.M. Sachs, Anthropometric manikin for acoustic research, J. Acoust. Soc. Am. 58 (1) (1975) 214–222.
[4] B. Gardner, K. Matin, HRTF measurements of a KEMAR dummy-head microphone, MIT Media Lab, http://sound.media.mit.edu/KEMAR.html.
[5] M. Tohyama, T. Koike, Fundamentals of Acoustic Signal Processing, Academic Press, 1998.
[6] M.R. Schroeder, D. Gottlob, K.F. Siebrasse, Comparative study of European concert halls: correlation of subjective preference with geometric acoustic parameters, J. Acoust. Soc. Am. 56 (4) (1974) 1195–1201.
[7] R.K. Cook, R.V. Waterhouse, R.D. Berendt, S. Edelman, M.C. Thompson Jr., Measurement of correlation coefficients in reverberant sound fields, J. Acoust. Soc. Am. 27 (6) (1955) 1072–1077.
[8] M. Tohyama, Sound and Signals, Springer, 2011.
[9] K.J. Ebeling, Statistical properties of random wave fields, in: W.P. Mason, R.N. Thurston (Eds.), Physical Acoustics, XVII, Academic Press, 1984, pp. 233–310.
[10] S.M. Baxter, C.L. Morfey, Angular Distribution Analysis in Acoustics, Springer, Heidelberg, 1986.
[11] A. Jeffrey, Handbook of Mathematical Formulas and Integrals, 3rd edition, Elsevier Academic Press, 2004.
[12] M.R. Schroeder, New method of measuring of reverberation time, J. Acoust. Soc. Am. 37 (3) (1965) 409–412.
[13] H. Yanagawa, Y. Yamasaki, T. Itow, Effect of transient signal length on cross-correlation functions in a room, J. Acoust. Soc. Am. 84 (5) (1988) 1728–1733.

[14] M. Tohyama, A. Suzuki, Inter-aural cross-correlation coefficients in stereo reproduced sound fields, J. Acoust. Soc. Am. 85 (2) (1989) 780–786.

[15] M. Tohyama, A. Suzuki, S. Yoshikawa, Correlation coefficients in a rectangular reverberant room, Acustica 39 (1) (1977) 51–53.

[16] M. Tohyama, A. Suzuki, S. Yoshikawa, Correlation coefficients in a rectangular reverberant room—experimental results, Acustica 42 (3) (1979) 184–186.

[17] A. Suzuki, M. Tohyama, Interaural cross-correlation coefficients of KEMAR head and torso simulator (in Japanese), IEICE Japan Technical Report EA80-78, Feb. 24 1981.

[18] J. Blauert, Spatial Hearing, MIT Press, 1995.

[19] Y. Ando, Auditory and Visual Sensation, Springer, 2009.

[20] Y. Ando, Signal Processing in Auditory Neuroscience, Temporal and Spatial Features of Sound and Speech, Elsevier, 2018.

[21] H. Yanagawa, T. Anazawa, T. Itow, Interaural correlation coefficients and their relation to the perception of subjective diffuseness, Acustica 71 (1990) 230–232.

[22] P. Damaske, Subjective Untersuchung von Schalfeldern, Acustica 19 (1967/68) 199–213.

[23] P. Damaske, Y. Ando, Interaural crosscorrelation for multichannel loudspeaker reproduction, Acustica 27 (4) (1972) 232–238.

[24] K.J. Gabriel, H.S. Colburn, Interaural correlation discrimination I. Bandwidth and level dependence, J. Acoust. Soc. Am. 69 (5) (1981) 1394–1401.

[25] B.C.J. Moore, An Introduction to the Psychology of Hearing, Academic Press, 1997.

[26] A. Suzuki, M. Tohyama, Interaural cross-correlation coefficients of KEMAR head and torso simulator in a free field (in Japanese), IEICE Japan Technical Report EA81-7, May 26 1981.

[27] G.F. Kuhn, Model for the interaural time differences in the azimuthal plane, J. Acoust. Soc. Am. 62 (1) (1977) 157–167.

[28] A. Suzuki, M. Tohyama, Interaural cross-correlation coefficients in stereo reproduced sound fields (in Japanese with English abstract), NTT R and D 36 (5) (1987) 681–690.

[29] M. Tohyama, A. Suzuki, Interaural cross-correlation coefficients in reproduced sound fields (in Japanese with English abstract), Architectural Acoustics Meeting Report AA81-27, Acoust. Soc. Japan, September 21 1981.

[30] H. Nomura, M. Tohyama, T. Houtgast, Loudspeaker arrays for improving speech intelligibility in a reverberant space, J. Audio Eng. Soc. 39 (5) (1991) 338–343.

[31] Y. Hirata, On the perception of acoustic space in reproduced sound fields (in Japanese with English abstract), Technical Report, EA80-71, IEICE Japan, 1981.

[32] R. Thiele, Richtungsverteilung und Zeitfolge der Schallrueckwuerfe in Raeumen, Acustica 3 (1953) 291–302.

[33] T.J. Schlutz, Acoustics of the concert hall, IEEE Spectr. (June 1965) 56–67.

[34] Y. Hirata, Reverberation time of listening room and the definition of reproduced sound, Acustica 41 (3) (1978) 222–224.

[35] T. Muraoka, T. Nakazato, Estimation of multi-channel sound field decomposition utilizing frequency-dependent inter-aural cross correlation (FIACC), J. Audio Eng. Soc. 55 (4) (2007) 236–256.

[36] Y. Takahashi, M. Tohyama, Multi-channel recording and reproduction for minimizing the difference in the spatial covariances between the original and reproduced sound fields, in: 19th International Congress on Acoustics, RBA-15-012, 2007.

[37] Y. Takahashi, A. Ando, Down-mixing of multi-channel audio for sound field reproduction based on spatial covariance, Appl. Acoust. 71 (12) (2010) 1177–1184, https://doi.org/10.1016/j.apacoust.2010.08.002.

[38] H. Hagiwara, Y. Takahashi, K. Miyoshi, Wave front reconstruction using the spatial covariance matrices method, J. Audio Eng. Soc. 60 (12) (2012) 1038–1050.

[39] Y. Hirata, Improving stereo at l.f., in: Wireless World, 1983 October, pp. 60–62.

[40] A. Suzuki, M. Miyoshi, M. Tohyama, Interaural cross-correlation coefficients and discrimination in reproduced sound fields (in Japanese), Technical Meeting, Hearing, H-85-7, Acoust. Soc. Japan, January 21, 1985.

[41] M. Kazama, S. Gotoh, M. Tohyama, T. Houtgast, On the significance of phase in the short term Fourier spectrum for speech intelligibility, J. Acoust. Soc. Am. 127 (3) (2010) 1432–1439.
[42] R. Meddis, L. O'Mard, A unitary model of pitch perception, J. Acoust. Soc. Am. 102 (3) (1997) 1811–1820.
[43] M. Tohyama, A. Suzuki, Inter-aural cross-correlation coefficients in stereophonic reproduced sound fields (in Japanese), JAS J. (April 1985) 3–13.
[44] H. Haas, Ueber den Einfluss eines Einfachechos auf die Heorsamkeit von Sprache, Acustica 1 (1951) 49–58.
[45] H. Haas, The influence of a single echo on the audibility of speech (translated into English), J. Audio Eng. Soc. 20 (2) (March 1972) 146–159.
[46] R.Y. Litovsky, H.S. Colburn, W.A. Yost, The precedence effect, J. Acoust. Soc. Am. 106 (4) (1999) 1633–1654.
[47] K. Terada, M. Tohyama, T. Houtgast, The effect of envelope or carrier delays on the precedence effect, Acustica 91 (6) (2005) 1016–1019.
[48] K. Terada, Signal dynamics and precedence effect on sound image localization and control (in Japanese), PhD Thesis, Kogakuin University, Tokyo, Japan, 2005.
[49] K. Terada, M. Tohyama, H. Yanagawa, T. Adachi, Precedence effect and signal envelope (in Japanese with English abstract), Technical Report of IEICE, Japan, EA99-109, 2000-03.
[50] K. Terada, M. Tohyama, The relationship between modulation depth and echo threshold on the precedence effect (in Japanese), J. IEICE Jpn. J 89-A (6) (2006) 494–501.

Index

A

B

C

D

G

H

I

J

K

L

M

Q

R

S

T

Printed in the United States
By Bookmasters